Palgrave Philosophy Today

Series Editor

Vittorio Bufacchi, Philosophy, University College Cork Philosophy, Cork, Ireland

T0253165

The *Palgrave Philosophy Today* series provides concise introductions to all the major areas of philosophy currently being taught in philosophy departments around the world. Each book gives a state-of-the-art informed assessment of a key area of philosophical study. In addition, each title in the series offers a distinct interpretation from an outstanding scholar who is closely involved with current work in the field. Books in the series provide students and teachers with not only a succinct introduction to the topic, with the essential information necessary to understand it and the literature being discussed, but also a demanding and engaging entry into the subject.

Rani Lill Anjum · Elena Rocca

Philosophy of Science

Rani Lill Anjum
Faculty of Environmental Sciences
and Natural Resource Management
and School of Economics and Business
Norwegian University of Life Sciences
Ås, Norway

Elena Rocca
Department of Life Sciences and Health
Oslo Metropolitan University
Oslo, Norway

ISSN 2947-9339   ISSN 2947-9347 (electronic)
Palgrave Philosophy Today
ISBN 978-3-031-56048-4   ISBN 978-3-031-56049-1 (eBook)
https://doi.org/10.1007/978-3-031-56049-1

Cover illustration: Photograph 51 by Rosalind Franklin and Raymond Gosling, Science History Images/ Alamy Stock Photo

This Palgrave Macmillan imprint is published by the registered company Springer Nature Switzerland AG
The registered company address is: Gewerbestrasse 11, 6330 Cham, Switzerland

If disposing of this product, please recycle the paper.

To our students

Preface

Back in 2020 we had an idea for a new philosophy of science course. The course was inspired by our joint work with engaging researchers, practitioners, and students in critical reflections about the philosophical foundation of science. Specifically, we have discussed how research methods, norms, and practices are motivated by philosophical assumptions that often remain implicit, what we call *philosophical bias in science*. Since our university offered almost exclusively interdisciplinary programs, students would encounter lecturers with diverse backgrounds and equally diverse perspectives. Disagreement could be over what counts as the best scientific methods, which results are more trustworthy, how to interpret and use scientific results, and what would be the best course of action given the available evidence. By teaching the students how to identify a range of philosophical assumptions, they were given some tools to understand and tackle the disagreement in a constructive way. Specifically, they learned to analyse scientific controversies where the disagreement among experts is not over the empirical facts, but how to interpret and evaluate those facts.

We wanted to write a book that is relevant for both philosophers and scientists, written by a philosopher and a scientist. The content is based on our individual and joint teaching and research. With this book, students can learn what we teach in our philosophy of science courses, and teachers can use it to develop their own courses. We hope to show the reader how one can apply and use philosophy to analyse and disentangle real cases of expert disagreement, but also to understand how scientific consensus requires some degree of philosophical consensus. In this respect, the book offers an introduction to philosophy of science that focuses on the application of philosophy to science.

This book has benefitted from insightful discussions and invaluable feedback from students, colleagues, and reviewers. We are grateful to the series editor, Vittorio Bufacchi, for giving us this opportunity and for encouragement and support along the way. We are indebted to the two anonymous reviewers for their constructive suggestions that significantly improved the quality of the book. A special note of thanks goes to our students at the Norwegian University of Life Sciences, who are the inspiration behind this book and to whom it is dedicated.

Ås, Norway   Rani Lill Anjum
                                                                        Elena Rocca

Contents

List of Figures

Part I
What's So Special About Science? Defining Science

Chapter 1
What Counts as Scientific Knowledge?

Trust and Distrust in Science ·

In September 2019, when 16-year-old climate activist Greta Thunberg addressed the US Congress on climate change, she brought the Intergovernmental Panel on Climate Change (IPCC) report on global warming as her testimony. 'I don't want you to listen to me. I want you to listen to the scientists', she said, referring to the broad scientific consensus behind the report. If the vast majority of scientists agree about the causes and effects of climate change, then why would anyone rationally distrust them?

When faced with urgent global challenges such as climate change, pandemics, refugee crisis, or loss of biodiversity, we look to science for answers. More often than not, however, we find that experts disagree. Such disagreement might be about explanations, predictions, solutions, or even how to frame a problem. How should we respond to this? Some might conclude that scientists don't really have the required knowledge to fix a problem. That might be the case, since scientific knowledge typically evolves over decades or centuries. Another response to expert disagreement is to deny that there was a problem in the first place. Climate sceptics are saying that there is no climate change to worry about, and during the Coronavirus Disease 2019 (COVID-19) pandemic, some even denied that the corona virus is real.

In times of broad scientific consensus, one might interpret any criticism of it as support of its radical opposition. If experts of medicine safety report that a new and much-needed vaccine can give some rare but serious adverse effects, one might (wrongly) interpret this as a support of the anti-vaxxer movement. Likewise, a critical argument against the scientific methods used in evidence-based medicine might be taken as an open invitation to quacks and healers. For similar reasons, philosophers of science are sometimes accused of being anti-science when questioning some deep-rooted assumption in scientific theory or method.

There are, however, better ways to respond to scientific disagreement and criticism. If given the right tools, expert disagreement can be made more transparent, better diagnosed, and responded to with rational arguments. This is what we want

R. L. Anjum and E. Rocca, *Philosophy of Science*, Palgrave Philosophy Today,
https://doi.org/10.1007/978-3-031-56049-1_1

to offer in this book, and to provide such tools will be our main focus in parts II and III. By becoming aware of some of the deeper sources of expert disagreement in science, we hope to make clear why it's not possible, and perhaps not even a rational goal, that all scientists and researchers should agree entirely. After all, what we take to be sound scientific knowledge today was once controversial, and some of our best current knowledge might be refuted or replaced in the future. By silencing opposition, or ignoring disagreement, we might prevent important scientific insights and progress.

In this part I, 'What's so special about science? Defining science', we give a general introduction to some central themes within classical and contemporary philosophy of science. The overarching question will be how to define science and scientific knowledge, and we consider a diverse range of answers offered by philosophers. Is science defined by its methods, by the scientific community itself, or is it a matter of power structures? Note, however, that none of the issues in philosophy of science are settled or generally agreed upon among philosophers or scientists. This might be frustrating for some readers, but the upside is that everyone can feel free to consider the matters for themselves, without feeling obliged to adopt one specific view. We therefore encourage our readers to keep an open mind and let us lead the way into the deep waters of philosophy of science. Hopefully, you will return safely to the surface by the end of the book, feeling wiser and intellectually refreshed.

What Type of Knowledge Deserves the Name 'Science'?

We all have some idea of what science is and what it means to do science. The question is whether there is a common definition of science that everyone could or should accept. Before we get deeper into the philosophy of science, we should therefore ask: What is science? What are the alternatives to science, and what separates science from these alternatives? Can there be one perfect definition of science, or is science many different things? It might help if we start with the very basics; with the concept itself.

The term 'science' is often reserved for the natural sciences, such as physics, chemistry, biology, and astronomy, thus excluding social science and the humanities. One might, however, argue that science is a much broader endeavour that should include for instance sociology, anthropology, history, economics, and linguistics. This would mean that most of our academic research areas are included. What about philosophy and literature? Or mathematics and statistics? None of these seems typical of science. Philosophy asks more questions than it can ever hope to answer, and statistics seems better fit as a tool for science. Still, we might be justified to include more than the natural sciences. In Latin, 'scientia' simply means knowledge, which is also reflected in the German 'Wissenschaft', that literally translates to 'knowledge-ship' or 'knowledge-making'.

This suggests that science is at least what we in philosophy call an epistemological matter, which comes from the Greek word for knowledge: 'episteme'. Episteme is the highest form of knowledge, and epistemology concerns itself with how such knowledge can be gained. Different answers have been offered in the history of philosophy, but we can broadly divide them into three categories: rationalism, empiricism, and perspectivism. These philosophical positions motivate different views on what type of knowledge we should aim for in science.

Rationalism: True Knowledge Comes from Reason

An early rationalist was Plato, a student of Socrates. Both Plato and Socrates argued that knowledge should be gained through reason. By using our rational capacity, we should seek to abstract some general principles from the messy and ever-changing reality that we experience through our senses. The highest form of knowledge would be the truths that hold universally, eternally, and ideally. This type of knowledge, Plato argued, could not be achieved by observing the material world, where we only find particular things that exist momentarily, in a particular time and place.

For Plato, the information that we can get through our senses will never represent true knowledge. He spoke of four levels of knowledge, where the lowest one comes from persuasion through rhetoric and second-hand sources (see Fig. 1.1). We might feel convinced by what we learn from teachers, politicians, or experts, but this would not give us deep knowledge of the true nature of things. Instead, it can be compared with an image, or a painting of reality. Certainly, the reality that an image is supposed to imitate would be more real than the image of it. If we instead investigate the material world for ourselves, by observing particular things and events directly, we can gain a higher level of knowledge. We wouldn't have to rely on authority but could check for ourselves how things are. This would bring us to the level of strong belief, or 'doxa'. At both these lower levels, we are investigating the material, particular world, using our senses. It is what we call empirical knowledge, and this does not represent true knowledge, or episteme, in Plato's rationalist philosophy.

TYPES OF EXISTENCE (ONTOLOGY)	TYPES OF KNOWLEDGE (EPISTEMOLOGY)		
UNIVERSAL TRUTHS AND PRINCIPLES	KNOWLEDGE VIA REASON AND ABSTRACTION	EPISTEME	LEVEL OF KNOWLEDGE
GENERAL CONCEPTS AND HYPOTHESES, MATHEMATICS	KNOWLEDGE VIA REASON AND ABSTRACTION		
MATERIAL OBJECTS	KNOWLEDGE VIA SENSES	DOXA	
IMAGES, ART, RHETORIC	SECOND-HAND KNOWLEDGE		

Fig. 1.1 Four levels of knowledge, from the lowest to the highest

In order to achieve the higher level, of true knowledge, we must instead use our reason to abstract general hypotheses, theories, and truths from what we observe in the material world. Being general, these truths are not available to our senses, but something we can only reach through reasoning. Scientific knowledge, as well as mathematical knowledge, will according to Plato be found on this level. The highest level of knowledge will be universal and ideal, and even explain the origin of what we observe in the material world.

Does this sound familiar for contemporary science? At least, it seems to fit some sciences. Plato looked to mathematics, where all truths are abstract, universal, and eternal. Mathematical truths are not found in time and space, which is also why we cannot observe them through our senses. What then with astronomy and theoretical physics, where one seeks to abstract some general principles and ultimately arrive at the laws of nature? While natural processes typically happen in open systems and with lots of contextual interferers, making them notoriously unpredictable, the scientist could derive such universal laws via abstract models, assuming closed systems and ideal conditions. This seems a good description of scientific theories about the behaviour of planets, atoms, and particles. But what about biology, ecology, and meteorology? Surely, these sciences must operate within the messy and particular features of reality, or?

Say we got a data set from a particular scientific experiment looking at *Drosophila melanogaster*, fruit flies typically used in research. These won't be the kind of fruit flies that we find in the kitchen late in the summer, but are flies that are lab-bred and genetically selected to be identical. They will be kept in strict isolation from the influences of anything that can affect their system or otherwise interfere with the experiment. The lab fruit flies are therefore thought of as models, and their experimental role is to represent the ideal fruit fly in a closed, isolated system and with perfect twin controls for comparison. Just as Plato suggests, the biologist is of course not primarily interested in these individual flies. They are instead used as tools to arrive at some general hypothesis or theory that would apply to *all* fruit flies, and preferably find something that can eventually be relevant also for human biology. This is crucial, since if a scientific theory only applies to a particular data set, the knowledge becomes useless and irrelevant for anything else. Once those individual fruit flies die, we have no use for the knowledge that is unique to them. The same can perhaps be said for linguistics, political science, medicine, and literature theory. As a field of research, one might aim for the more general principles, trends, patterns, and classifications that can be used to describe, understand, explain, and predict beyond what can be observed at a particular time and place. From a rationalist ideal of knowledge, it seems that social anthropology and history would not be good candidates for sciences, at least not if they describe specific contexts without attempting to abstract or generalise anything from them.

For the rationalist, the highest form of knowledge would be those principles or laws that have the widest application. When astronomer Cecilia Payne-Gaposchkin found evidence that stars are composed primarily of a single element, hydrogen, it revolutionised the field of astronomy and motivated a whole new scientific area, of astrophysics. Likewise, we consider Isaac Newton's law of universal gravitation and

Charles Darwin's theory of evolution by natural selection to be some of the most important scientific theories ever proposed. If Newton's laws only applied to cannon balls, and Darwin's theory only applied to a specific strand of fruit flies, they would lose much of their scientific value.

The rationalist assumption that the highest form of scientific knowledge lies in generalisations and universal laws, has been challenged in recent years. In *How the Laws of Physics Lie*, philosopher Nancy Cartwright argues that the knowledge represented by such laws has no application outside of abstract and idealised models, or at best describes some highly artificial experimental settings. In her terminology, the scientific models and tools designed to create law-like behaviour are nomological machines ('nomos' meaning 'law'), suggesting that the laws are made up more than they describe reality. 'For the fundamental laws of physics do not describe true facts about reality. Rendered as descriptions of facts, they are false; amended to be true, they lose their fundamental explanatory power' (Cartwright, 1983, p. 54). If we take the idealised model to be more important than the reality that it is supposed to explain, then science has missed out on practical application, relevance, and explanation. Any description or prediction would then only hold true within the model itself. Cartwright urges that scientists and philosophers of science pay more attention to the particular and context-specific aspects of reality, rather than to the general and ideal. This is not an attractive option for the rationalist perspective, since one would then miss out on the most valuable form of knowledge, which are the abstract, eternal, and universal truths.

Empiricism: True Knowledge Comes from Sense Experience

A very different take on knowledge comes from the empiricists, who said that we should only trust as true knowledge what we can get from observation, or experience ('empeiría' in Greek). Using our senses, we seem to have more or less direct access to the material world. Before we start having sense experiences, our minds would be like blank slates, or empty buckets, ready to be filled with empirical knowledge. David Hume, who was an empiricist, said that any attempt to theorise beyond what we have observed should be considered metaphysical speculations. Metaphysics is a branch of philosophy where one asks about the true nature of reality, independently of what we can prove or justify empirically. For an empiricist, this would not amount to knowledge. Instead, science should rely heavily on observation and experimentation.

A famous definition of science comes from logical positivism, or logical empiricism. This is a strict form of empiricism, applied to the domain of science. The movement developed in the early twentieth century among scientists and philosophers who, like Hume, stated that there are only two types of trustworthy knowledge: that derived through logic and analysis, such as mathematics, and that derived empirically from sense experience. Any scientific knowledge, they thought, should be empirical. In addition, the logical empiricists stipulated that any scientific claim must be empirically testable and verifiable. As a minimum, then, it should be possible to specify the criteria under which the scientific claim would be empirically confirmed. If no such verification criteria could be offered, one is not dealing with a candidate for scientific knowledge, but metaphysical speculation, or simply nonsense.

Most scientific theories will have some empirical element to them that can be explicated and tested. But what about some highly abstract physical theories, stipulating the existence of multiple dimensions and parallel universes? How could such theories be empirically verifiable? Perhaps they would be closer to metaphysical speculation? While the rationalist might appreciate the abstract, general, and ideal nature of such a theory, it wouldn't really fit the typical empiricist approach to science. According to the logical empiricists, one should stick to the empirical data throughout the whole scientific process: when generating hypotheses, testing them, and reporting the results. One should avoid introducing theoretical explanations or conclusions that go beyond the empirical evidence.

A scientist committed to empiricism would need to find ways to translate unobservable aspects of their theories into something observable and testable. In research design, this part of the scientific process is the operationalisation: to make explicit what operation we would have to perform in order to empirically identify a phenomenon that we want to study. One might say that distance should be identified by a measuring rod, for instance, and time should be identified by a clock. According to physicist and Nobel laureate Percy W. Bridgman, however, one must be careful that different operationalisations for the same phenomenon are not conflated or treated as the same. It matters, for instance, whether one measures distance using a rod or light years. These two operations might not even measure the same phenomenon. The solution, Bridgman suggests, is to specify what exactly it is that we are measuring, to avoid any confusion about the phenomenon that we identify. 'To say that a certain star is 10^5 light years distant is actually and conceptually an entirely different *kind* of thing from saying that a certain goal post is 100 meters distant' (Bridgman, 1927, pp. 17–18).

Let's consider an example. In psychology and cognitive science, one typically assumes that people have a rich inner life, of thoughts and feelings, even though one cannot access these from direct observation. What we know about other people's inner lives is primarily based on what they say and how they behave, assuming that they don't lie or hide their feelings. In philosophy, behaviourism is the empiricist idea that human behaviour should be explained without making a reference to unobservable entities. Any emotion or beliefs must then be observed through someone's behaviour. One might talk in terms of stimulus and response, or input and output. Another empiricist approach is to measure observable brain activity such as electric impulses or bloodstream. This requires advance measuring tools, and a question in philosophy of science is what gets lost of the phenomenon in the process of operationalisation. For instance, one might ask whether a feeling or thought is the same as a certain electric impulse in the brain, or something else entirely.

An example of operationalisation in cognitive science, is to measure cognitive response to a stimulus by using a functional magnetic resonance imaging (fMRI) scanner. This is an instrument that registers the changes in blood flow that corresponds to brain activity. Using an fMRI scanner, for instance, one might observe which parts of the brain light up when a person is looking at a series of photographs evoking specific emotions, thus linking a certain emotion to the activity of specific brain regions. In what has been known as 'the dead salmon study', Craig Bennett and colleagues got statistically significant responses when running a test of their experiment on a fresh, but dead fish from the supermarket (Bennett et al., 2010). Their conclusion was not that the fish had emotional responses to the images, but that such false positive result must be avoided by improving the statistical methods. Another conclusion could be that the operationalisation itself is flawed. What one can observe by using an fMRI scanner, namely statistically significant changes associated with blood flow, might be different from the phenomenon that one wants to study, which is the emotional response. For strict empiricists such as the logical empiricists, failure to translate a theory into something observable would be a serious problem. Unless the process of operationalisation maintains the essence of the phenomenon, one cannot test or verify the theory. Given their strict definition of scientific knowledge, an untestable theory should be discarded as meaningless and unscientific.

The logical empiricists have significantly influenced how we consider the scope and limits of science. Their verification criterion separates scientific claims from both religious and philosophical claims. Neither of these disciplines makes claims that can be empirically tested or verified, which is why the logical empiricists say they are both pointless, and even meaningless. Another consequence of the verification criterion is that one must separate clearly between claims about empirical facts and claims about moral values. Since there is no way to empirically verify that it is wrong to inflict pain on animals, for instance, it is not something that we can meaningfully claim according to the logical empiricist.

The fact-value distinction between *what is* and *what ought to be*, was central to Hume's empiricist framework. Normative claims about what we ought to do, he said, can only be justified by referring to a further normative premise, which ultimately comes from ourselves. Say we accept that animal testing inflicts pain. This is not sufficient to derive that we ought to stop animal testing. For this, we need an extra premise, saying that one should not inflict pain. Without this normative premise, the conclusion does not follow. There is no fact in the world to decide whether it is true or false that we ought to stop animal testing. To mix facts and values is also known as the naturalistic fallacy, stating that it is logically invalid to derive what *ought* to be the case from what *is* the case. Many scientists and philosophers accept the distinction between facts and values, and that science should stay clear of value claims. Still, the idea that science is or could be entirely objective, factual, and value-free has been challenged by many contemporary philosophers, including Sandra Harding, Donna Haraway, Helen Longino, Heather Douglas, Uma Narayan, Evelyn Fox Keller, Ruth Anna Putnam, and Cordelia Fine. One objection is that even to say that science ought to be value-free is to make a normative claim, which itself is a value judgement.

Fig. 1.2 Duck or rabbit? A
matter of perspective
(Wikimedia Commons)

Fig. 1.2 Duck or rabbit? A matter of perspective (Wikimedia Commons)

Perspectivism: All Knowledge Is Situated

Perspecitivism is a third position within epistemology, which states that knowledge always comes from a certain perspective. This theory is neither purely rationalist nor strictly empiricist. Instead of seeing knowledge as something that is gained from only sense experience or only reason, knowledge requires both. What all perspectivists oppose, is the idea that knowledge can ever be 'a view from nowhere'. Rather, there is always a perspective involved: 'a view from somewhere'. Consider the famous duck-rabbit illustration, where the same object can be observed either as a duck or as a rabbit (Fig. 1.2). There are many such examples of ambiguous figures, where most people who view them see only one of the alternatives while others immediately see the other. Perspectivism is the idea that all empirically based knowledge will be subject to ambiguity and interpretation. In a scientific setting, this means that two scientists looking at the same data can arrive at widely different descriptions, explanations, and conclusions. Using the duck-rabbit figure, we can see how the same observation can lead to different theories, since starting from the observation of a bird would inspire a whole different scientific story than if one observes a mammal.

It might seem far-fetched that there could ever be such an ambiguity in a scientific setting, but history of science is full of examples where the same empirical observation is taken to support different theoretical explanations. The most famous example is from astronomy and the development of various Earth-centred (geocentric) and sun-centred (heliocentric) models of the universe, which lasted for thousands of years, from the Babylonians 4000 BC and far into the seventeenth Century. During all this time, scientists and philosophers had found the empirical evidence to support their model. In philosophy of science, this is the problem of underdetermination; that a set of observations cannot settle, or *determine*, which theoretical explanation is the correct one. This means that for any data set, there could be several theories that are able to explain them equally well. Which we take to be most plausible typically depends on what other theories we hold true.

Perspectivism has its philosophical influence from Immanuel Kant, who argued that the empiricist and the rationalist ideals are both untenable. Although knowledge requires sense experience, our minds are not just blank slates or empty buckets that we fill up with knowledge as we get more experience. Reason plays an important part, by organising our sense experience into certain categories that are essential to our human perspective. Kant was concerned with our common human perspective and not with cultural or individual perspectives. We know that humans perceive the world differently than dogs, for instance, since dogs can hear and smell things that we cannot, while they cannot see all the colours that humans can. This necessarily affects the type of sensory information one can access. In the same way, Kant would say, there are ways of thinking that we cannot avoid as humans, and that might be shared with other animals: that we see the world in objects, in causal relationships, and as happening in time and space.

With Kant, there is a move away from the purely empiricist idea that we access the external world directly, without any filter, towards acknowledging that all data must necessarily be processed via our human senses and cognitive capacities. We can recognise this way of thinking in phenomenology, which focuses less on a mind-independent reality and more on the way things appear to us: as phenomena. The idea that even scientific knowledge depends on perspective is often met with criticism, especially from those who think that science is valuable only insofar as it's independent, unbiased, and objective. According to Kant, a mind-independent reality would be inaccessible to us and remain purely speculative.

In this sense, Kant and the logical empiricists agree that empirical knowledge should be free of metaphysical speculation. Unlike the logical empiricists, however, Kant does not see metaphysics, ethics, or religion as meaningless, but as different types of claims altogether. Philosophy seeks knowledge that is neither analytical nor empirical. In stating that 'science must be empirically testable and verifiable', or that 'science must be value-free', philosophers are making informative, yet non-empirical claims. In contrast to empirical claims, which are typically stated about a specific time and place, philosophical claims are stated as if they were universal and necessary. In Kant's own words, these claims are 'synthetic a priori'. They are informative (synthetic), so one can disagree over whether they are true or false, yet non-empirical (a priori), since no empirical facts about the world could ever settle such disagreement. In parts II and III, we will see that many scientific controversies involve exactly this type of non-empirical disagreements, which is why they must be settled by using arguments, and not only by producing more scientific evidence.

If we accept that even scientific knowledge is mind-dependent and relative to some perspective, doesn't this mean that anything goes and that all perspectives are equally valid? Can we meaningfully say that science uncovers facts, or are all scientific truths constructions of our own minds? Does perspectivism mean that science has a subjective element, where something can be true for one person but false for another? That depends. In philosophy, perspectivism comes in a wide spectrum of alternatives.

One realist version of perspectivism, promoted by philosopher of physics Michela Massimi, is *perspectival realism*. This is a theory that emphasises how the academic-institutional perspective influences and shapes science. All scientific inquiry is seen as the product of historically and culturally situated communities, where knowledge is generated via certain resources, tools, instruments, and practices. Science is then not objective in the sense 'mind-independent', but is a perspective that includes scientific theories, evidence, and principles of knowledge set by the scientific community.

Standpoint theory is another type of perspectivist theory, with roots in Marxism. The theory has been developed by feminist philosophers of science such as Sandra Harding and Donna Haraway and comes in different versions, some more relativist than others. Broadly speaking, standpoint theory sees all knowledge as influenced by socio-political context, such as gender, class, ethnicity, race, and physical capacities. Our experiences and social perspectives shape and limit what we can know, which types of scientific questions we ask, and the knowledge-gaps that we notice. Everyone will have their own blind spots; what we are unable to see simply because it has never concerned us. When perspectives are shared by scientists, they will not even recognise their own blind spots, but wrongly consider their views to be neutral and objective.

A more relativist version of perspectivism is found in Thomas Kuhn's theory of *scientific paradigms*. According to him, normal science is done within a wider scientific framework where relevant theories, tools, methods, concepts, and problems are decided by the scientific community. Some might be critical of this description of science because it suggests that all paradigms or perspectives are equally valid. Unless there are some additional, overarching criteria for saying that one scientific paradigm is better than another, Kuhn's concept of a paradigm makes scientific knowledge entirely relative to a specific scientific framework and community.

Constructivism is a form of perspectivism that sees all knowledge as mental or social products. Science does not simply uncover facts about the world, but produces new facts that are shaped by us and our own interests. The way we classify things, for instance, might depend on what we see as relevant similarities and differences, but this does not guarantee that such classifications exist in the world. Similarly, our scientific theories contain human-made concepts and categorisations, and these might not reflect the physical reality. Constructivism, interpreted as a relativist position, will according to its critics make it difficult to even talk about scientific truth or knowledge, since there is nothing to know outside our own mental or social constructs. This form of constructivism is thus anti-realist about scientific truths.

We see from these examples that perspectivism can be understood very broadly, as referring to a common, human perspective, or it can be understood in a more radical relativist way, as subjectivism, where all knowledge is relative to the individual perspective. Common for all these versions, however, is that they reject any version of scientific realism that demands that we access the external world directly, neutrally, and independently, what in philosophy is referred to as 'direct', 'commonsense', or 'naïve' realism.

Concluding Remarks

Still today, the philosophical traditions of rationalism, empiricism, and perspectivism continue to tacitly influence science, by giving rise to very different scientific traditions and approaches. Some scientific traditions place more emphasis on theory and use of abstract models and idealisations, in line with rationalism. Other traditions are more empiricist and see theories as something that is only credible insofar as they are backed up by, or generated from, empirical data. Perspectivism is common in the social sciences, where many are sceptical toward the ideal of mind-independent, neutral, and universal truths. This means that although scientists might not engage extensively or explicitly in the philosophical debate on knowledge, their choice of theory, method, and practice could reveal that they have already sided with a philosophical position without even knowing it. By learning about philosophy, and philosophy of science, one can become aware of the philosophical foundations of science and understand better what motivated the norms and research practices that one takes for granted.

Not everybody agrees that philosophy of science is useful for scientists. Physicist Richard Feynman, for instance, was critical of philosophy, and the famous joke that 'philosophy of science is as useful to scientists as ornithology is to birds' has been attributed to him on several occasions. Similarly, physicist Stephen Hawking opens his book *The Grand Design* by stating that philosophy is dead and that scientists have taken over as the bearers of the torch of discovery in our quest for knowledge. In contrast, philosopher Daniel Dennett has said that 'there is no such thing as philosophy-free science, just science that has been conducted without any consideration of its underlying philosophical assumptions' (Dennett, 2013, p. 20).

This book is written with two aims in mind, which also reveals our own preferred view in philosophy of science. First, we want to explain why science cannot avoid engaging with philosophy, implicitly or explicitly. Second, we want to demonstrate how implicit philosophical assumptions, when made explicit, can be used as an analytic tool for making scientific disagreement and controversy more transparent for all stakeholders of science. Our hope is to create a common basis for constructive, transdisciplinary discourse.

Chapter Summary

In this first chapter, we started from the idea that science is about generating knowledge. We then looked at three quite different philosophical takes on what knowledge is: rationalism, empiricism, and perspectivism. We explained how each of these starting points would give different answers to the question of what counts as scientific knowledge. An important question remains: How can we separate knowledge in general from scientific knowledge? Exactly what type of knowledge should count as scientific? Certainly, not all types of knowledge should meet the criteria, since one can know many things that are not considered scientific knowledge. One might for instance *know* the name of the woman who invented insulin (Dorothy Hodgkin), *explain* why the train is delayed (flawed signal), and *reliably predict* that we will get wet from the rain while waiting outside (heavy rain and no cover). Certainly, none of these insights deserves a place in a science journal, even though they are both empirical and verifiable. What, then, is it about scientific knowledge that makes it so special? We usually take for granted that science is by far the best way to gain knowledge, but why, and how exactly does it do so? Are some scientific approaches better than others, and if so, how can we decide what these are? What are the alternatives to science? To answer these questions, we need to move on with our quest to define science. As should have become clearer, however, no consensus has yet been reached on the matter among philosophers or scientists.

Further Introductory Reading

There are many books on the market that offer an introduction to philosophy of science. The shortest one, *Philosophy of Science—A Very Short Introduction*, by Samir Okasha (2002), is easy to read and covers an impressive list of topics in such small space. More in-dept books on the topic include *What Is This Thing Called Science*, by David Chalmers (2013). *An Introduction to the Philosophy of Science* by Lisa Bortolotti (2008) is an excellent teaching resource with exercises and discussion questions for each chapter. It also has a glossary and a detailed thematic bibliography. A more recent book with the same title is by Kent Staley (2014). While all these books focus almost exclusively on the intellectual works of men, some classic books within feminist philosophy of science include *Science as Social Knowledge: Values and Objectivity in Scientific Inquiry* by Helen Longino (1990) and *Whose Science? Whose Knowledge?* by Sandra Harding (1991). The Cambridge Elements in Philosophy of Science series includes a number of short and accessible texts, and new books are made available for free download in the first two weeks after publication.

Further Advanced Reading

More comprehensive works include *The Routledge Handbook of Feminist Philosophy of Science*, edited by Sharon Crasnow and Kristen Intemann (2021), and *The Routledge Companion to the Philosophy of Science*, edited by Statis Psillos and Martin Curd (2014). These are both up-to-date anthologies that cover a wide range of themes, written by experts for students and other newcomers to the field. If one wants to learn more about the philosophical debate on rationalism, empiricism, and perspectivism as presented here, we recommend Nancy Cartwright's classic texts *How the Laws of Physics Lie* (1983) and *The Dappled World* (1999). Here she challenges the rationalists' agenda in science to search for the most universal and general truths and ignore more relevant and applied forms of knowledge for our messy reality. Another relevant resource is Michela Massimi's (2022) *Perspectival Realism,* which is a contemporary version of perspectivism applied to scientific knowledge. Massimi has also edited the textbook *Philosophy and the Sciences for Everyone* (2015).

Free Internet Resources

The School of Philosophy, Psychology and Language Sciences at the University of Edinburgh offers several free online courses, including an introduction to philosophy of science course developed by Massimi and colleagues: *Philosophy and the Sciences. Examining Philosophy's Relationship with the Physical and Cognitive Sciences.* On the course website, one can find videos, handouts, and a recommended textbook.

Study Questions

1. Do you think expert disagreement is a problem for public trust in science? Explain your answer.
2. How should we respond to critical voices in scientific discourse? Could there be ways to recognise good or bad criticism?
3. What does it mean to say that science is an epistemological matter? What does epistemology mean?
4. What is rationalism? Do you see any signs of Plato's ideals of knowledge in science today or is this way of thinking outdated?
5. What was Cartwright's criticism of the rationalists? Do you agree with her?
6. In what ways do empiricists disagree with rationalism? Which of these two theories of knowledge do you find most attractive and why?
7. What were the two criteria for knowledge stated by the logical empiricists? Do you think this is a good way to define science? Could there be knowledge that meets the criteria but doesn't count as scientific?
8. What is the fact-value distinction and why is it so important to the empiricists? What is the criticism of this distinction, and do you agree with it?
9. Broadly speaking, how do you understand perspectivism about knowledge? Which version do you prefer and why?
10. What do you think about the claim that it can be useful for scientists and others to learn about philosophical theories of knowledge?

Sample Essay Questions

1. Present two or more philosophical theories of knowledge and discuss their impact on scientific thinking. Which of these has influenced your own understanding of science and how?
2. How do you see the role of science with respect to (a) truth, (b) value claims, or (c) social representation? Use what you have learned about logical empiricism (positivism) and perspectivism to support your arguments.

3. Discuss the tension between the need for scientific consensus and the challenge of expert disagreement. For instance, should science be trusted only if there is a majority consensus on the matter? If so, how do you see science being challenged and pushed forward? Do you think there is a danger of distrust in science when experts disagree?

References

Bennett, C. M., Baird, A. A., Miller, M. B., & Wolford, G. L. (2010). Neural correlates of inter-species perspective taking in the post-mortem Atlantic salmon: An argument for proper multiple comparisons correction. *Journal of Serendipitous and Unexpected Results, 1*, 1–5.

Bortolotti, L. (2008). *An introduction to the philosophy of science.* Polity Press.

Bridgman, P. W. (1927). *The logic of modern physics.* Macmillan.

Cartwright, N. (1983). *How the laws of physics lie.* Clarendon Press.

Cartwright, N. (1999). *The dappled world: A study of the boundaries of science.* Cambridge University Press.

Chalmers, A. F. (2013). *What is this thing called science?* Hackett Publishing.

Crasnow, S., & Intemann, K. (Eds.). (2021). *The Routledge handbook of feminist philosophy of science.* Routledge.

Curd, M., & Psillos, S. (Eds.). (2014). *The Routledge companion to philosophy of science.* Routledge.

Dennett, D. (2013). *Intuition pumps and other tools for thinking.* W. W. Norton.

Harding, S. (1991). *Whose science? Whose knowledge? Thinking from women's lives.* Cornell University Press.

Longino, H. (1990). *Science as social knowledge: Values and objectivity in scientific inquiry.* Princeton University Press.

Massimi, M. (Ed.). (2015). *Philosophy and the sciences for everyone.* Routledge.

Massimi, M. (2022). *Perspectival realism.* Oxford University Press.

Okasha, S. (2002). *Philosophy of science: A very short introduction.* Oxford University Press.

Staley, K. (2014). *An introduction to the philosophy of science.* Cambridge University Press.

Chapter 2
Should Science be Defined by Its Methodology?

What Is the Best Method?

Continuing the investigation of how to separate scientific knowledge from other types of knowledge, one suggestion is that science is restricted by its methods. There is a significant difference, one might think, between making a claim that just happens to be true and making a scientifically grounded claim. We might have good reasons to believe that it will rain tomorrow, for instance that it has been raining for several days already, but a meteorologist who arrives at the same conclusion will have done so using scientific methods. A psychic might be right in their prediction about the weather and the meteorologist wrong, but this doesn't make the true prediction more scientific than the false prediction. One view, therefore, is that scientific knowledge counts as such only if it is produced using the correct systematic approach. This might sound strange, since it suggests that we would trust a false scientific claim more than a true unscientific claim. Why on earth would we do that?

A reason for placing so much trust in methods is that history of science has shown us that truth is an elusive matter. Scientific truths come and go, and scientific knowledge for which we had decisive evidence at some point, was later refuted or replaced. How, then, can we know that our best current knowledge won't also be replaced in the future? Returning to the question, *What is science?*, one suggestion is therefore that science is defined by its methods. All knowledge that is generated with the right method would then count as scientific, and otherwise not. But this brings us to another question: *Which methods?* Are some methods better than others? If so, what makes them better? Is there a single perfect method, or do we need a combination of methods? These are ongoing discussions within philosophy of science. Other philosophical discussions are more specific. How important is it, for instance, to use experiments, models, or controls in science? And what are the advantages and disadvantages of studying a complex phenomenon by separating and isolating individual elements? Are scientific tools neutral, or do they introduce extra-evidential assumptions into the study?

© The Author(s), under exclusive license to Springer Nature Switzerland AG 2024
R. L. Anjum and E. Rocca, *Philosophy of Science*, Palgrave Philosophy Today,
https://doi.org/10.1007/978-3-031-56049-1_2

In science, one typically sees methods as the specific techniques, tools, and procedures used to set up and carry out a study. In philosophy of science, in contrast, one is primarily concerned with investigating some general features and principles that many methods have in common and see how these can be justified. This is a *methodological* discussion. Today, for instance, many take for granted that experimentation is the best method for generating knowledge and evidence, at least in the natural sciences and medicine. This was not always the case. When the experimental method was first developed and used by prominent scientists such as Ibn al-Haytham and Galileo Galilei, it was controversial because Aristotle never used physical experiments or mathematical models (Gower, 1997, p. 23; Tbakhi & Amr, 2007). Eventually, however, this became the default scientific approach. We will here present two broad, systematic approaches that are commonly used for advancing and establishing scientific knowledge: the inductive method and the hypothetical-deductive method. Before going into details, we must first explain what it means that a method relies on induction or deduction.

Deductive and Inductive Reasoning in Science

Aristotle introduced two types of reasoning, deductive, and inductive, as part of his system of logic. Deduction is when we start with a general premise and use this to draw conclusions about a single case. This is a top-down approach, where the direction of reasoning goes from the general to the specific. Say we know that (1) *all* philosophers are mortal, and that (2) Hypatia was a philosopher. Based on these two premises alone, we can conclude that Hypatia was mortal. In deductive arguments there is no new information in the conclusion than what we already stated in the premises, so the conclusion might seem a bit obvious. Applied to scientific reasoning, say we know that (1) all heated metal expands (general premise) and that (2) we are heating up a piece of metal (specific premise). From this, we can logically deduce that this metal will expand (specific conclusion), which follows directly from the two premises.

A Deductive Argument

General premise	All As are Bs
Specific premise	x is A
Specific conclusion	x is B

This is only one type of deductive inference, but common for them all is that there shouldn't be more information in the conclusion than what was assumed in the premises. Philosophers have argued that deductive inferences can be useful for both predictions and explanations. We might use the premises to predict the conclusion (that the piece of metal will expand), or to explain the conclusion (because all heated metal expands, and this piece of metal was heated). This is the deductive-nomological model of explanation, and 'nomos' here refers to the general law in the first premise. A question is how we arrive at these general premises in the first place. One way to do this is through the other type of reasoning, which Aristotle called induction.

Induction is when we start from specific premises and use these to arrive at a general conclusion. This is a bottom-up approach, where we start from specific observations with the aim of arriving at a general theory. For instance, how do we know that all heated metal expands? What we have observed is that some, or even many, heated pieces of metal expanded, so the conclusion that all metal behaves like this certainly goes beyond the empirical evidence reported in the premises. To draw a conclusion about *all* instances based on observation of *some* instances is typical for inductive inferences. While deductive arguments contain no more information in the conclusion than what we already stated in the premises, inductive arguments always conclude beyond the premises. In this sense, inductive conclusions provide new information.

An Inductive Argument

Specific premise	Some As are Bs
General conclusion	All As are Bs

David Hume was critical of inductive inferences and argued that they violate both the rules of logic and the empiricist criterion of knowledge. If we can only know what we have empirical evidence of, then any conclusion about unobserved events will be logically invalid. All predictions about future events based on past observations are inductive, and therefore unwarranted. Inductive reasoning is nevertheless common in science and is used whenever we need to infer (1) from some instances to all instances, (2) from the observed to the unobserved, or (3) from the past to the future. Since science often aims to find theories that hold generally, also for unobserved and future instances, the problem of induction represents a real challenge for scientists.

In contemporary science the problem of induction is taken seriously. When reporting their research findings, scientists are reluctant to make the inductive conclusion that the results apply generally or beyond the study. If a new medicine is tested on 100.000 individuals and found safe and effective, it would be an inductive, hence invalid, inference to conclude that it will be safe and effective also for those who were not in the study. Nevertheless, this is what happens whenever scientific results get applied in other contexts than those that generated them. And isn't that an important aim of research? If we couldn't use the knowledge generated from research for technological innovation, public policy, or any implementation of change, then what would be the value of the knowledge for society? We might of course say that knowledge is valuable in itself, regardless of practical application. Nonetheless, we will need inductive reasoning to get from observed data to general theory. To allow some form of induction therefore seems vital, both for science and its application.

The Inductive Method: How to Get General Theories from Unbiased Observations

Empiricist philosopher Francis Bacon saw science as valuable for society precisely because it can be used for technological innovations to improve our lives. This motivated him to develop an inductive method. Bacon called the method *The New Scientific Instrument*, which is the title of his book from 1620: *Novum Organum*. His method is inductive because it begins with examining the particulars and ends in general knowledge. According to this approach, scientific inquiry ought to start from objective, and systematic observations, collected under a variety of conditions, without any prior judgement or expectation as to what one will find. Bacon was critical of the influence of Aristotle, which had dominated science and philosophy for centuries with aggressive support of the church. In his lifetime, prominent scientists had been killed or threatened to silence for publicly challenging the Aristotelian geocentric worldview. Among these were Giordano Bruno, who was burned to death in 1600 after seven years of arrest, and Galilei, who spent the last 9 years of his life in house arrest, after being forced to retract his views twice, in 1616 and 1633. Bacon's new method was developed to ground science on empirical facts alone, thus liberating science from the dogmatic influences of authority, religion, and Aristotelian or other forms of metaphysics. In separating science from anything non-empirical, Bacon inspired the logical empiricists in dismissing all metaphysical speculation or value judgement from the domain of science.

Bacon's method takes us from observation of particular events to the most general theories, in a gradual process. First, one should make many and varied observations, under different conditions. It is important here that one does not anticipate the findings or allow the observations to be influenced by prejudice, bias, or priority. Observations should be done in a systematic and critical manner, and Bacon suggests that one organises the empirical data into tables to compare them and look for patterns. One should take notice if something is always present, always absent, or always increases or decreases with the same degree in all the observed instances (Bacon, 1620, Book II, p. 15). This first step then gives rise to inductive inferences of more general axioms and hypotheses. By recording both positive and negative instances, one can better choose between possible explanatory hypotheses and eliminate all those that have counterexamples. Throughout this process, the generation of new axioms and hypotheses must be done stepwise, to avoid making hasty and unsupported generalisations that conclude beyond the empirical evidence. One must be careful that the inductive inferences don't over-stretch, or that conclusions get transferred to contexts where they don't apply. The human intellect, he says, 'should be supplied not with wings but rather weighed down with lead, to keep it from leaping and flying' (Bacon, 1620, Book I, p. 104). The final step is to test the most general hypothesis, which is a candidate for a law of nature, and see whether it gets confirmed or refuted. If confirmed, we have successfully produced and verified a general theory based on empirical facts alone.

Bacon was one of the first to discuss the problem of bias in science, which is when our expectations, prejudices, dogmas, or interests influence the research without our awareness. His work includes a detailed discussion about how the scientific method could fail to produce reliable results because of human error. Instead of ignoring these weaknesses of the human intellect, Bacon urges that one should be aware of them and try to minimise their impact on the scientific process. He identifies four types of biases, or 'idols', that tend to distort the human intellect in the scientific inquiry. The first type is *the idols of the tribe*, which comes from our human nature and perception. We tend to find regularity and order where there is none, and to seek confirmation of our preferred theories. We typically ignore empirical evidence that goes against our theories, even when it outweighs the supporting evidence. Instead, Bacon says, we should treat positive and negative evidence with the same attitude. The second type of bias is *the idols of the cave*, which come from our individual character, upbringing, and influences. Our personal interests or perspectives make us focus on certain fields of investigation and ignore others. Any enthusiasm towards one specific theory should therefore be treated with suspicion, to ensure that the scientific investigation remains balanced and critical. Bacon also warns against *the idols of the marketplace* and *the idols of the theatre*. The marketplace refers to the meaning of words and concepts, and how use of language enables and hinders our thinking. When a distinction is made in language, he says, we start to think that it reflects an important division in nature, which then makes it difficult to change our thinking. The idols of the theatre are when we dogmatically follow truth from authority, which for Bacon included Aristotelian metaphysics and Plato's rationalism, but also the empiricist error of generalising from a few observations to a theory. Finally, he warns against superstition and religion.

To sum up, the inductive method consists of four broad steps. (1) Make many and unbiased observations, (2) find patterns of regularity, (3) construct general hypotheses, and (4) test and results. If the test confirms the theory, it is verified, and if not, the theory is falsified.

Induction in Science: The Case of Thalidomide

We can illustrate the inductive method with an historical example of a scientific discovery that revolutionised medicine. Until the 1960s, drug manufacturers globally were not provided with an explicit or detailed method to prove the safety of their products. This lack of regulation led to distribution of several harmful chemicals. A famous example is thalidomide, a medicine that was found to cause serious foetal deformities (for historical details, see Dally, 1998). When thalidomide was first marketed in 1957 in over 40 countries, it was not only commercialised as a sedative, but also used as an antiemetic, to treat morning-sickness in pregnant women. In 1961, Australian obstetrician William McBride reported to the medical journal *The Lancet* that he had observed an increased number of congenital abnormalities in association with maternal use of thalidomide. In particular, he had seen several children born with limb and ear impairment, a very rare medical condition which had become inexplicably more frequent in that period. McBride expressed concern and advanced the hypothesis that thalidomide was the cause of foetal malformations. The same hypothesis was advanced independently by the German paediatrician Widukind Lenz, who investigated the gestational details of mothers of deformed children compared to mothers of healthy children and found out that the first were often thalidomide users. Based on these observations and on accumulating case reports, the drug was withdrawn from the market (Vogel, 1995).

After this, national and international surveys showed that thalidomide had caused severe birth defects in over 10.000 children between 1957 and 1962. In 1964, a report sponsored by the UK government showed systematic observations supporting that almost any organ could be affected by the drug, and that damage was primarily seen in the limbs. Evidence that thalidomide causes birth defects is today overwhelming; different species of experimental animals are susceptible to the effect at different degrees, and birth defects have been observed recently in Brazil, where the drug has been re-introduced as therapy for leprosy. The general hypothesis is now seen as verified, and the discovery also led to new theoretical insights about the permeability of the placenta to different types of chemicals. Until the 1960s, medical students were taught that the placenta offers a perfect protection for the foetus. This belief had been held for centuries in the medical community despite available knowledge that some maternal conditions and behaviours, such as rubella virus infection and alcohol consumption, could affect foetal development.

The devastating effects of thalidomide is one of the biggest scandals in the history of medicine. Since this discovery, the methods for assessing medicines for safety have changed radically. Today we take for granted that one needs clinical data to evaluate risk and safety of various human exposures to chemicals *before* approving a new medicine. Indeed, there are national and international agencies responsible for monitoring side-effects of medicines after they have been introduced on the market. Because of the thalidomide disaster and the discovery that medicines might harm the foetus, most new medicines today are not even tested on or marketed to pregnant women.

Recall that a main motivation for Bacon to make space for inductive inferences in science was precisely that science and technology could be used to advance society and improve our lives. Without the tool of induction, he thought, science becomes impotent, stripped of its power and value for society. In the case of thalidomide, for example, the decision to withdraw it from the market was based on the inductive inference that what was observed in some instances could happen to more children in the future. Other conclusions based on induction are the general causal mechanisms used to explain the foetal malformation and the reasoning that led to a change in risk assessment procedures of new medicines.

The Hypothetical-Deductive Method: A Problem-Driven Approach

Bacon's method was an attempt to make inductive inferences as empirically justified as they could possibly be. Still, he cannot escape Hume's problem of induction, that the general theory goes beyond the empirical evidence. Philosopher of science Karl Popper responded to the problem of induction in a radical way. To avoid making unwarranted claims, we should never conclude that a general hypothesis or theory is proven true, or *verified*. Popper agreed with Hume that no amount of single observations could ever be sufficient to verify a general hypothesis. Instead, he said, a single observation would be enough to refute it. For instance, we might think that *all* swans are white because we have only observed white swans. But once a black swan was observed, the hypothesis that all swans are white was falsified. If we think that all use of thalidomide is effective and safe, and it turns out that it is actually harmful in some cases when used during pregnancy, then that hypothesis must also be falsified. Perhaps, then, one could have a scientific method that resists induction by only allowing negative results as conclusive?

Popper was critical of research programs that seemed primarily concerned with confirming their theories by only looking for supporting data and dismissing any negative evidence. He had noticed that some contemporary theories were too adaptable to be refuted through observation. Instead of taking apparent counterexamples as a sign of weakness, these theories allowed the counterexamples to be explained away by referring to the theory itself. According to Popper, both Marxism and Freud's psychoanalytic psychology were cases of pseudo-science because they had some self-confirming elements that made them immune to falsification. This was no way to do science, he thought. Instead of gathering lots of data to confirm our theories, Popper says we should look for data that would refute them. Only then will we see how strong the theory is. Even if we cannot verify a hypothesis in this way, we can eliminate false ideas and increase our confidence in those theories that are not (yet) falsified. Only the fittest theories will survive. The hypothetical-deductive method should according to Popper be used to put our hypotheses through hard testing and see if they resist falsification. The method is deductive because any positive result must be taken as tentative rather than conclusive. If so, we can say that the theory is supported, or *corroborated*, by the empirical evidence, but it might still be falsified by future observations.

Another common criticism against Bacon's method, apart from it being inductive, is that one cannot start from observations without any specific focus. How is it possible to collect data if we don't even know where to begin? Shouldn't any scientific study start from a theory or some problem that one wants to solve? Bacon himself used the example of studying the nature of heat, and he suggested that one collects all relevant observations for that purpose to look for patterns of similarities, differences, exception cases, limit cases, and so on. These data would then be used to formulate hypotheses and design experiments for testing them. This suggest that one needs some initial focus to guide our data collection, especially to know which data are relevant and which are not. In the hypothetical-deductive method, one therefore starts with formulating a problem. The problem is what should be explained, and hopefully solved. Once there is a problem, one might also have some hypothesis or theory that could explain it. This is the stage of creative speculation, and Popper says it doesn't matter how one arrived at a specific hypothesis. An important step, however, is to find a suitable way to test the hypothesis empirically. For this, one must think in the following way: If the hypothesis is true, then what would be an observable consequence of it? Is it possible to design a study or experiment to see whether this empirical consequence follows or not? If so, which other supporting hypotheses do we need for the test to be a good one for this specific purpose? Only after these steps are carefully made can the actual test be performed.

Note, however, that the reliability of the results we get from using the test depends on more than one main hypothesis. Although an experiment is designed to test the main hypothesis, we also need to consider whether we have designed a suitable test for it, including any supporting hypotheses. What we test is therefore not the main hypothesis directly, but a testable implication of the hypothesis. But this is not all: together with the hypothesis and its implication, the test involves a whole set of assumptions, some of which are part of the test itself. Say we use a mathematical model, an algorithm, or some lab instruments. Once we accept the results they give, we have also accepted the tools as reliable and suitable for the purpose. In Chapter 1, we mentioned the dead salmon experiment, where one assumed that the results from using an fMRI scanner could reliably be interpreted as brain activity. When the dead salmon gave positive results, it challenged the trustworthiness of the whole experimental setup, including the commonly accepted methodology. While Popper took for granted that a negative result of the hypothetical-deductive method meant that the main hypothesis was falsified, others have argued that one can never know exactly which of the assumptions were wrong. This has been referred to as the problem of holism in theory testing, with Pierre Duhem, Willard Quine, and Mary Hesse as some of the main critics. If all theories come in a network of interconnected assumptions, they argue, one cannot test any of them in isolation. All one can conclude from a negative result, is that at least one of the assumptions was wrong. A problem with this is that we risk falsifying valuable theories based on one false premise.

The hypothetical-deductive method was proposed as a superior alternative to the inductive method, but the focus in the two methods is quite different. Bacon explained in detail how to generate a theory from unbiased observations, which requires systematic inductive reasoning. Popper and Carl Hempel both saw this as a non-starter, so the hypothetical-deductive method includes no step for how to arrive at a general hypothesis or theory. Instead, their focus is on how to *justify* hypotheses. For instance, the hypothetical-deductive method includes the extra step of how to translate the theory into something suitable for empirical testing. In philosophy of science, some say that the hypothetical-deductive method is a shift from the context of *discovery* to the context of *justification*.

We can summarise the hypothetical-deductive method in five main steps. (1) Find a problem, (2) suggest a hypothesis of what is the source of the problem, (3) deduce some hypothesis implications that can be tested empirically, (4) design an appropriate test, and (5) test and results. If the test confirms the theory, it is supported (but never verified), and if not, the theory is falsified.

Hypothetical-Deduction in Science: The First Vaccination

An historical example following the steps of the hypothetical-deductive method is Edward Jenner's experiment to test vaccination as a way to confer immunisation against illnesses (for historical details, see Riedel, 2005). Jenner was a natural scientist who lived in eighteenth century England. At that time, smallpox, an illness that had devastated Europe since the Middle Ages, caused 400.000 deaths a year and led to blindness in a third of the survivors. It was therefore a societal problem to which medicine and natural science devoted considerable effort.

When Jenner formulated his hypothesis in 1779, there was no known mechanism for the development of the disease, because the germ theory of disease would be introduced almost a century later. Nevertheless, Jenner had a good deal of knowledge available about the illness and the medical practices in use at the time, and from this background knowledge he built his hypothesis. First, it was common knowledge that smallpox survivors were immune to the illness for the rest of their lives. Second, it was known that such immunity could be transmitted by inserting, or *inoculating*, infected material from smallpox pustules of a sick person into a healthy person. The practice of inoculation was first introduced in Europe from Asia thanks to the efforts and influence of Mary Wortley Montague, who had suffered the illness herself and lost her brother to it. At the Ottoman court in Istanbul, Montague had witnessed the medical procedure of immunising healthy people against smallpox by infecting them with small amounts of material from lesions taken from sick patients. She continued to advocate for the adoption of this procedure in England, and even had her young children inoculated as early as in 1718 and 1721. Later in Russia, Catherine the Great famously led a ground-breaking public health campaign to promote inoculation against public resistance (Ward, 2022). Professional inoculation procedures became a big scale practice all over the Western world, and statistical surveys were used to demonstrate a 10 times lower death rate in inoculated people compared to people who got smallpox naturally. Nevertheless, this practice was risky and 2–3% of inoculated patients died from smallpox, transmitted it to healthy individuals, or contracted a secondary disease.

In addition to the medical knowledge of the time, Jenner had heard all his life that dairymaids were immune to smallpox after having suffered from cowpox, a similar disease that was dangerous for cows but not for humans. Jenner hypothesised that the immunity given by cowpox might be transmittable to people by inoculation, and therefore used as a safer strategy for protection against smallpox. The testable implication of such hypothesis was that it might be possible to immunise the body against smallpox by deliberately inoculating the cowpox disease. In 1796 he carried out an experiment to test his hypothesis implication. Jenner inoculated an 8-years-old boy with material from cowpox lesions, causing a slight fever and discomfort for a couple of days. After two months, he inoculated the boy with smallpox, and no disease developed. Jenner concluded that the hypotheses was supported: cowpox gave immunity to smallpox, and such immunity could be transmitted by inoculation.

Since Jenner's hypothesis was initially met with scepticism by the medical community, he turned to the task of generating more evidence to corroborate it. A second testable implication of his hypothesis is that, at population level, it might be statistically visible that persons who suffered from cowpox are resistant to smallpox, or do not develop symptoms after inoculation with smallpox. He tested and confirmed this implication by conducting a national survey in 1799. Following Jenner's experiments and observations, the use of his new practice to confer immunity, which he called 'vaccination' (from the Latin name for cowpox, *Variola Vaccinae*), spread rapidly in and outside of England. Since the modern version of germ theory of disease got accepted and one gained more understanding about the body's immune system, the theory of transmittable immunity has survived and continues to evolve.

Can Science be Defined by Falsifiability?

Against the logical empiricists, and their verification criterion of scientific knowledge, Popper proposed the falsification criterion: *A theory is scientific only if it could in principle be falsified from a single observation.* Falsifiability is according to Popper the demarcation of science. It is what separates science from non-science, or what he called pseudo-science. 'Pseudo' means 'fake' or 'imitation', so pseudo-science would have the appearance of science, but not meet the criteria of being self-critical and empirically falsifiable. Today, we are mainly concerned with pseudo-sciences that promote what we take to be empirically ungrounded theories. We might think that climate change deniers are getting support from pseudo-scientific publications, for instance, or that any evidence claimed to support that crystals can heal us must be bogus. Most important for Popper, however, is the idea that science ought to be a self-critical activity, where the scientific community does not protect their own theories, but systematically works to test and challenge them. If a theory cannot even in principle be falsified by data, it does not deserve the name science.

What exactly does it mean to be falsifiable in principle? And how easy is it to spot the difference between a falsifiable and unfalsifiable claim? Is it true that all aspects of a scientific theory must be falsifiable? Let us consider which of the following claims could be proven wrong by an observed counterexample.

- There is a 5% chance of rain tomorrow.
- This treatment is 80% effective.
- All ravens are black.
- All whales are mammals.
- All heated metal expands.
- All electrons are negatively charged.
- All humans are mortal.

- All women are biologically programmed to want children.
- All actions are ultimately selfish.

The first two claims are probabilistic predictions that refer to the outcome being likely or have a certain chance of happening. This makes it difficult to falsify the claims based on a single observation, since the estimate might have been accurate even if it fails to predict what actually happened. If the estimate says that there is a miniscule chance of rain tomorrow, it might still be correct even if one wakes up to pouring rain. The rest of the claims are general, about *all* instances of a class, which satisfy Popper's ideal that science makes strong and bold claims. Still, they are not all easy to falsify empirically. Some of them might be stipulating a definition or a classification. All whales are mammals, because of some biological properties that all mammals share. One might then argue that if one ever observed a whale that wasn't a mammal, then that thing could not really be a whale. This suggests that the claim is not an empirical one, suitable for falsification. Similarly, if the claim that all electrons are negatively charged is stated as a definition or stipulation, there is no point in testing it empirically. The same argument then goes for the claim that all humans are mortal. If one thinks that any immortal being could not be human, one assumes that humans are essentially and necessarily mortal. If instead interpreted as an empirical claim, it would be falsifiable. It might happen that scientists one day succeed in curing humankind of mortality by altering the biological mechanism of aging. Other claims on the list might belong to the core of a theory and be a starting point for research rather than what one wants to investigate. That all actions are ultimately selfish, for instance, is difficult to refute as an empirical claim. Some argue that altruistic actions are only apparently so, because we act selflessly only because it makes us feel good. If so, then our actions are selfish after all. Evolutionary psychology and sociobiology are both research programs that start from the assumption that human behaviour and interactions can be explained from certain biological principles, as argued by Richard Dawkins in his famous book *The Selfish Gene.* One might thus not be interested primarily in testing the core theory, but to see how much the theory can explain.

All this suggests that whether we take a claim to be in principle falsifiable, partly depends on what role that claim has in the theory. This is why it is important to be transparent about whether a statement is made as an empirical claim, a scientific result, or stipulated as a premise. Without such transparency, something might appear to be an empirical finding that is in fact a theoretical assumption.

Do We Need Plural Methods?

Bacon, Popper, and the logical empiricists all argued for a single scientific method that should be applied universally. But is it reasonable to demand that there is only one correct method, and that if the method isn't applied, the results cannot be trusted as scientific? We have looked at two methods, and both were able to generate valuable scientific knowledge. How should we understand this? Is there really such a clear distinction between inductive and deductive reasoning, where only one is scientific? Some philosophers have argued that it shouldn't matter which methods one uses. In *Against Method* Paul Feyerabend argued that when it comes to method, history of science shows us that almost 'anything goes'. According to him, we should not search for a single correct scientific approach, to be applied universally. Even Bacon admitted that many scientific discoveries were made entirely by accident. Perhaps we need a plurality of scientific methods?

Chapter Summary

We have seen that some scientific inquiries might start from a problem or hypothesis, while others might start from gathering data in a systematic way to see which hypotheses can be made. The hypothetical-deductive method falls under the first approach, while the second approach is inductive. Instead of seeing these as conflicting methods, we could accept Feyerabend's idea that different methods can lead to valuable results. If so, inductive and deductive reasoning seem complementary and useful for different stages of the scientific process. One might have more than one hypothesis to explain a problem, for instance, and then use some form of inductive reasoning to help choose which to test. There are, however, some problems with the pluralist position. One is that different methods sometimes generate conflicting results. Which results we should trust will then depend on which of the methods we think are better for the purpose. This is an issue that we will discuss further in parts II and III. Another problem is how we decide which methods to accept and why. Is it enough that they might generate scientific knowledge, or should there be some further quality check done by the scientific community? What if the community is critical of a new approach and doesn't acknowledge it as scientific? The role of a peer-community for defining science will be the topic of our next chapter.

Further Introductory Reading

Most introductions to philosophy of science books will include chapters on scientific methods, including the inductive method, falsification, the hypothetical-deductive method, and methodological pluralism. One of the best books for this is *What Is This Thing Called Science*, by David Chalmers (2013). A book that is primarily about methods is *Scientific Method: An Historical and Philosophical Introduction* by Barry Gower (1997), which offers a nice overview of this topic. A short introduction to the inductive and deductive approaches can be found in 'Theories of scientific method' by Nancy Cartwright, Stathis Psillos, and Hasok Chang (2003). For an introduction to the problem of holism in theory testing used against Popper's falsificationism, Milena

Ivanova (2021) has written a book in the Cambridge Element series for Philosophy of Science: *Duhem and Holism.*

Further Advanced Reading

Francis Bacon's (1620) *Novum Organum* is easy to read and offers insights that are still relevant and valuable for scientists. Other classic texts are Karl Popper's (1959) *The Logic of Scientific Discovery* and Paul Feyerabend's (1975b) *Against Method.* For the more advanced reader, we also recommend Mary Hesse's (1974) *The Structure of Scientific Inference.*

Free Internet Resources

A favourite of ours is the classic lecture by Imre Lakatos (1973), 'Science and pseudoscience', which is freely available as both podcast and transcript. The full text of Bacon's (1620) *Novum Organum* can be found on Early Modern Texts. *Stanford Encyclopedia of Philosophy* has an entry on 'Scientific method', written by Brian Hepburn and Hanne Andersen (2021).

Study Questions

1. What do you think about the idea that there is only one correct scientific approach?
2. What is deductive reasoning? Would you agree that science can rely on deductive reasoning alone?
3. How is Bacon's method inductive? In what sense is the method empiricist?
4. Hume's problem of induction has been taken seriously by both philosophers and scientists. What were some responses to the problem?
5. What were some of Bacon's concerns about bias in science? Are any of these more common than others, you think, or more serious?
6. Some have said that Bacon's inductive method is a naïve form of inductivism, or even naïve empiricism. Do you agree? Explain.
7. What are some differences between the inductive method and the hypothetical-deductive method? Do you see any strengths or weaknesses with one or the other?
8. Which of the two methods do you prefer, if any, and why?
9. Do you think that different types of sciences would prefer different types of methods? If so, what could help us choose the best method for our research?

10. What are your thoughts on Popper's falsification criterion of science? And the idea that one can never verify but only falsify a scientific hypothesis?

Sample Essay Questions

1. Present and compare inductive and deductive reasoning and discuss how they influence scientific approaches. Which of these has influenced your own understanding of scientific methods and how?
2. How do you think that science should deal with verification and falsification of general theories and hypotheses? Use what you have learned about the problem of induction, the problem of falsification and the problem of theory holism to support your arguments.
3. Present the four biases, or idols, presented by Bacon. Try to find examples from your own discipline where any of these biases could affect the results of the research. Specifically, can you find some assumptions that most people in your discipline would accept, although they might be biases?

References

Bacon, F. (1620). *The new organon* (F. H. Anderson, Ed.). Bobbs-Merrill. A free version can be found on the Early Modern Texts webpage. https://www.earlymoderntexts.com/assets/pdfs/bacon1620.pdf

Cartwright, N., Psillos, S., & Chang, H. (2003). Theories of scientific method. In M. J. Nye (Ed.), *The Cambridge history of science* (vol. 5, pp. 21–35). *Modern physical and mathematical sciences.* Cambridge University Press.

Chalmers, A. F. (2013). *What is this thing called science?* Hackett Publishing.

Dally, A. (1998). Thalidomide: Was the tragedy preventable? *The Lancet, 351,* 1197–1199.

Feyerabend, P. (1975a). How to defend society against science. *Radical philosophy* (pp. 261–271). Stegosaurus Press.

Feyerabend, P. (1975b). *Against method.* New Left Books.

Gower, B. (1997). *Scientific method: An historical and philosophical introduction.* Routledge.

Hepburn, B., & Andersen, H. (2021). Scientific method. *Stanford encyclopedia of philosophy* (Summer 2021 ed.). E. N. Zalta (Ed.). https://plato.stanford.edu/archives/sum2021/entries/scientific-method

Hesse, M. (1974). *The structure of scientific inference.* Macmillan.

Ivanova, M. (2021). *Duhem and holism.* Cambridge University Press.

Lakatos, I. (1973). *Science and pseudoscience.* LSE podcast and transcript. https://www.lse.ac.uk/philosophy/science-and-pseudoscience-overview-and-transcript/

Popper, K. (1959). The propensity interpretation of probability. *British Journal of Philosophy of Science, 10,* 25–42.

Riedel, S. (2005). Edward Jenner and the history of smallpox and vaccination. *Proceedings* (Baylor University, Medical Center), *18,* 21–25.

Tbakhi, A., & Amr, S. S. (2007). Ibn Al-Haytham: Father of modern optics. *Annals of Saudi Medicine, 27,* 464–467. https://www.ncbi.nlm.nih.gov/pmc/articles/PMC6074172/

Vogel, F. (1995). Widukind Lenz. *European Journal of Human Genetics, 3*, 384–387. https://doi.org/10.1159/000472329

Ward, L. (2022). *The empress and the English doctor how Catherine the great defied a deadly virus.* Blackwell.

Chapter 3
Is Science Defined by Its Community?

The Scientist as a Lone Genius: A Myth?

We often hear stories of major scientific advances made by individual scientists, such as Charles Darwin's theory of evolution, Alexander Fleming's discovery of penicillin, or Benjamin Franklin's discovery of electricity. The narrative of the lone genius is also reflected in the science community, where recognitions and awards are often aimed at individual achievements. For instance, the highest form of recognition that a scientist can receive, the Nobel Prize, is reserved for 'the worthiest person', or 'the person who made the most important discovery' in their field. How common is it, though, that a revolutionary scientific discovery is made by an individual researcher, in isolation from a wider scientific community? To which extent does scientific advances depend on the community, and to which extent on the sagacity of a few talented individuals?

One famous story is Rita Levi-Montalcini's discovery of the neural growth factor, for which she won several prestigious science awards, including the Nobel Prize. Her work started during the Second World War, when fascist dictator Benito Mussolini banned all Jews from Italian universities and denied them the right to publish in science journals. Levi-Montalcini lost all access to laboratory facilities and had to seek refuge in the Italian countryside. Isolated from the scientific community, she installed some scientific equipment in her bedroom, including some home-made tools, and wandered around local farms in search of fertilised eggs to dissect. It was in this improvised lab that she made her early observations of the nervous system development in chicken embryos. In her autobiography, Levi-Montalcini writes that her interest in how neural growth is induced by the surrounding tissue first started when she read an article by a research group at Washington University in St. Louis. She and her assistant managed to publish their observations in an American journal in 1940, with results that conflicted with the original American experiments. After the war Levi-Montalcini was invited by the research group in Washington University

R. L. Anjum and E. Rocca, *Philosophy of Science*, Palgrave Philosophy Today, https://doi.org/10.1007/978-3-031-56049-1_3

in St. Louis, where she replicated the results from her home-lab experiment and continued her work to isolate the nerve growth factor in tumours.

What would have happened to Levi-Montalcini's work if she never got to publish her research, or didn't get to continue her work in an established lab or university? Could her work revolutionise medicine if it had remained an isolated effort in her bedroom? Probably not. Without her finding a way to replicate and communicate the results to an established scientific community, there could be no acknowledgement or influence no matter how important the findings. This has some radical consequences for how we think about science and research. First, it suggests that science is partly or wholly defined by a community. Second, it means that scientific discoveries are dependent on recognition from this community. Third, we might have to accept that science is a democratic matter, at least to some extent. That science is somehow relative to a community goes against the logical empiricists (or positivist) ideal of science as an objective and independent activity, where truths can be discovered or established based on empirical evidence alone. We will here look at different views on how science happens, or should happen, within a scientific community. We begin with presenting Thomas Kuhn's theory of scientific paradigms and how he sees science as a genuinely social practice.

Thomas Kuhn: 'Normal Science Happens Within a Paradigm'

In Chapter 2 we discussed the idea that science is defined by its methods, suggesting that a finding is scientific only insofar as it is generated using the right approach. A question remains, then, who should decide which methods count as scientific? One possible answer is that this is decided by the scientific community. When entering a scientific discipline, as students or researchers, we must accept much more than a certain methodology. We are also taught which scientific theories are the most important ones, who are the authorities in the field, and which problems remain to be solved. Some research questions are considered relevant and worthy of funding, while others are not, and there will be criteria for what types of data or results one should be seeking. Theoretical concepts will be clearly defined, methods described, tools standardised, and science journals and publishers will be ranked according to their status and relevance. These are the rules of the game that one must accept if one wants a degree or career in that field of study. Taken together, all this amounts to what historian of science Thomas Kuhn called a 'scientific paradigm'. In *The Structure of Scientific Revolutions*, he argues that science typically happens within this type of framework. In Kuhn's terms, 'normal science' consists in the joint effort of the scientific community to develop the theoretical details of the paradigm and solve any remaining problems of the theory. He compares normal science to puzzle-solving, where scientists work to fill in the knowledge-gaps: the missing pieces of the puzzle.

An example of normal science is the ambitious Human Genome Project, aiming to sequence and map all the genes of the human species (Sulston & Ferry, 2002). From 1990 to 2003, scientists worked toward a shared goal, based on a common set of theories, assumptions, tools, and techniques. One common theoretical framework, which worked also as a grounding motivation for the whole project, was the central dogma of molecular biology, stating that DNA codes for RNA, which codes for proteins, and proteins are responsible for physiological functions. This can be seen as a form of genetic determinism, which is the assumption that the traits of an organism are determined by its genetic makeup. Some scientists have referred the human genome as 'the book of life', claiming that it offers the complete set of instructions to create a human being.

Normal science, in Kuhn's version, is a stage of dogmatic acceptance of the foundational premises of the paradigm. Laws that play a central role in the paradigm are not seen as empirical claims to be tested, but instead as definitions, or stipulations. Kuhn even compares normal science to religion and says that the scientists must have strong faith in the main theory, with no serious intention to challenge or replace it. Progress in science does not happen by falsifying and dismissing bad theories, as Popper proposed. Instead, one leaves time and space to develop a theory in all its details and see how much it is able to explain. When encountering negative results or anomalies, there might be a plausible explanation for this that could help develop the theory further. An anomaly is when the observation does not fit the theoretical expectation, which means that the result is unexpected or unsupported by the theory. If one should falsify a theory whenever a piece of the puzzle doesn't fit, one might be missing out on important new insights. Typical for normal science, therefore, is that one responds to an anomaly with what Kuhn calls an 'ad hoc' hypothesis. These are hypotheses that are especially designed to explain why the results were not as one would expect. Perhaps some tools need to be improved or some of the previous observations or assumptions were wrong. Or it might be that parts of the theory need to be revised slightly.

In the Human Genome Project, one very surprising finding, or anomaly, was that humans have much fewer genes than expected given the complexity of the organism. Rather than the estimated 100.000 protein-coding genes, one could only find 26.000 to 40.000 genes, which isn't much considering that a fruit fly has around 14.000 protein-coding genes. An ad hoc hypothesis that was used to explain that the modest increase in number of genes between invertebrates and humans allowed for a dramatic increase of complexity, was that there isn't a one-to-one relationship between a human gene and a human trait. Another finding was that, just like in the genome of other species, also in the case of human genome there was no even distribution of protein-coding genes. Instead, coding-dense areas were separated by large repeated non-coding sequences. An ad hoc hypothesis used to explain the long areas of non-coding chromosomes (called 'junk-DNA' because it had no known function), was that they were remaining from evolution and used to have some important functions in the past.

Within a paradigm, Kuhn says, an ad hoc hypothesis plays the role of protecting the main theory and save it from falsification. Still, there is a limit to how many anomalies a paradigm can persist, and at some point there will be a paradigmatic crisis where there are simply too many counterexamples that cannot be explained. At this stage, one might even start questioning some of the more foundational assumptions of the paradigm, and philosophical debates often emerge as a result. The idea of 'junk DNA', for instance, is currently under debate, especially after the discovery that portions of such DNA are transcribed into RNA and cover important regulatory functions even though they don't code for any proteins. Scientists are debating how big of a portion of non-coding DNA lacks an actual function. Some have argued that the whole hypothesis of junk DNA has functioned as a deterrent for research programs to look for a possible function (Buehler, 2021).

A scientific revolution happens when one paradigm is replaced by another paradigm and the rules of the game change. Typically, Kuhn thought, this does not mean that scientists convert from one to the other, but that members of the old paradigm eventually die out and young scientists join the new paradigm instead. In the new paradigm, there will be other theories, authorities, research agendas, journals, knowledge gaps, and so on, that steer the scientific activity. The contributions of individual scientists are valued according to how important they are for the research agenda of the new paradigm. Discoveries or theories that didn't fit the old paradigm might see an increased interest in the scientific community. Since the community is also responsible for giving science awards, it could take years before a discovery is recognised as a major contribution. It might have been a paradigmatic discovery that radically changed the course of a discipline or gave rise to a new field of research. If so, there would be little enthusiasm in the scientific community when the discovery is first reported, and the peers might even deny that there is a discovery.

One paradigm changing contribution to genetics came from Barbara McClintock's genetic analysis of multi-coloured maize, where she discovered the phenomenon of *jumping genes* (Pray & Zhaurova, 2008). Until the late 1940s, the paradigm of genetic heredity assumed that the genome is made up by entities, the genes, arranged in a stable linear order in chromosomes. Stable mutations in genes would then be inherited by the offspring, and accumulating mutations would allow the evolution of the species by natural selection. This model, together with the idea of how complex genomes evolve, was first challenged and eventually modified by a series of discoveries. It all started when McClintock observed that some genes were capable of moving along the genome. She showed that such mobile elements, later named 'transposons', not only were able to change their positions within the chromosome, but also to alter the activation of other genes. McClintock's findings were initially met with resistance since it was not easy to explain them within the established paradigm of genetics at the time. Because of this, she waited to make her results public until 1951, after others had reproduced her findings. Decades later, in 1983, McClintock was awarded the Nobel Prize for her discovery. Her work is considered a significant break-through in genetic research, paving the way for the new field of epigenetics. Today it is commonly recognised that jumping genes are abundant in eukaryotes. They make up over 60% of the maize genome and even 40% of the human genome, although the majority of these are inactive. In the new current paradigm, the study of evolution and adaptation requires the mapping and understanding of transposable elements (Bourque et al., 2018; Ravindran, 2012).

What happens, then, when one paradigm is replaced with another? According to Kuhn, there is no genuine discussion or disagreement between two scientific paradigms, since the types of questions, theories, approaches, and concepts they deal with are simply too different. Two paradigms are *incommensurable*, meaning that they cannot be compared because they don't share the same measures or standards. Any attempt at constructive dialogue would amount to talking past each other, and the best way to understand the rise of a new paradigm is as a new beginning rather than as an improvement of the previous one. For instance, one might argue that the periodic table of chemical elements wasn't meant to develop or improve the Ancient four elements of earth, water, air, and fire, but simply made that discussion outdated and irrelevant. So even though both Aristotelian physics and particle physics are concerned with irreducible elements, the meaning and theoretical roles of these have radically changed. The new paradigm won't spend much time or energy arguing against Aristotelian physics, once properly established, and if they did try, they would end up talking past each other.

Some Responses to Kuhn

While Kuhn's social theory of science has been highly influential, especially within philosophy and social science, his ideas have been challenged and developed. Some oppose the relativist definition of science and reject the idea that scientific knowledge and progress is dependent on the dominant paradigm. Kuhn's theory suggests that a scientific discovery only counts as such once it is acknowledged by the members of the scientific community, or paradigm. If the scientific community decides which theories are true, what types of data are relevant, what counts as results, and how they should be interpreted, then does this also mean that empirical data are scientifically relevant only if they fit what else we already know? In the previous chapter, we mentioned the idea of Duhem, Quine, and Hesse, that theories include a whole network of knowledge, and that one cannot easily pick out one of them as the right candidate for falsification. The same can be said for knowledge within a scientific paradigm. One anomaly might not falsify a core theory, but it might indicate that some of the supporting hypotheses were wrong. In 'Science as social? - Yes and no' Susan Haack compares scientific activity with filling in a crossword puzzle, where new empirical evidence will be evaluated according to how well it fits with what we already know or else be dismissed:

> Experiential evidence is the analogue of the clues, background information of already-completed entries. How reasonable an entry in a crossword is depends on how well it is supported by the clue and any other already-completed, directly or indirectly intersecting, entries; how reasonable, independently of the entry in question, those other entries are; and how much of the crossword has been completed. (Haack, 1996, p. 80)

Haack denies that knowledge is socially constructed or nothing more than the product of consensus among members of a scientific community. That science is social, she says, does not mean that there are no objective standards for what counts as good empirical knowledge.

Even if we accept that scientific knowledge is not entirely relative to a paradigm, the idea that empirical evidence must fit into a coherent set of knowledge still raises some important questions. What knowledge might we miss out on if scientific findings must fit the already accepted assumptions of the paradigm? What if some or even many of those assumptions are wrong? How does one correct or challenge a whole paradigm if what counts as science is defined by the paradigm itself? A danger is that science becomes a self-confirming and conservative system, rather than a self-critical and progressive one. These are some concerns that one might raise against the idea that science is relative to a community.

Some philosophers agree with Kuhn's starting point, that scientific knowledge is a social matter, yet disagree with the idea that two paradigms don't compare or compete, and that no progress can happen from one paradigm to another. Imre Lakatos, for instance, accepts the social, historical, and normative aspect of scientific paradigms, and also the description of how the research community deals with anomalies by proposing ad hoc hypotheses. However, he argues that Kuhn's notion is too strict to include scientific progress and rationality across the different frameworks. Lakatos introduces the notion of research programmes and argues that these are better descriptions of how science actually works than the Kuhnian paradigms, and that they also allow us to make a rational choice of which framework to accept. Instead of simply being indoctrinated or bullied into a paradigm, Lakatos thought that the different research programs both compare and compete. They also enable scientific progress, which he says typically happens when two or more research programmes compete to make the best theoretical explanations and predictions. In any research programme, there will be a hard core of the theory, with a belt of auxiliary hypotheses to protect this hard core from falsification. The research program is progressive, he says, when it grows and successfully transforms counterexamples into supporting evidence for the theory, resulting in novel discoveries and predictions. In contrast, a research program is regressive, or degenerating, when the counterexamples pile up and require an increasing number of ad hoc hypotheses that do not improve the predictive power of the theory. Scientific revolutions thus happen when one research programme supersedes another, by overtaking it in progress. This, however, might change, and a degenerating research programme might recover and become progressive at a later point.

Another philosopher of science who offers an alternative to Kuhn's theory is Helen Longino. In *Science as Social Knowledge*, she develops a position that she calls 'critical contextual empiricism'. In line with Kuhn, Longino agrees that scientific reasoning and observation don't happen in a vacuum, and that science takes place in a social world. Scientific standards, methods, and values are set by and dependent on the scientific community, and it therefore matters, she says, *who* is included. If empirical evidence or discovery depends on the social, political, moral, and cultural context of the scientific community, it is crucial that the community is diverse and also includes marginalised groups. Michela Massimi agrees with Longino that scientific knowledge is historically and culturally situated. In *Perspectival Realism*, she is critical of Kuhn's theory because he assumes that scientific knowledge is produced by individual scientific communities, in disjoint 'silos'. To understand complex problems, she argues, one needs to go beyond disciplinary boundaries. Hence, the scientific community cannot be understood in the narrow sense of shared membership of a paradigm. Massimi is also concerned with the *reliability* of science, of how scientific knowledge gets justified. Such justification, she argues, is necessarily perspectival, and this is why we need a plurality of scientific perspectives to produce reliable scientific knowledge.

In the beginning of this chapter, we asked whether a scientific discovery could ever happen in a social vacuum, independently from a scientific community or peers. If we accept a social theory of science, as proposed by Kuhn, Lakatos, and Longino, among others, the answer would simply be 'no'. Even when a discovery is initially made in isolation, such as Levi-Montalcini's work in her home laboratory, it is only when the discovery takes part in the context of a community that it can contribute to knowledge production. We might in hindsight attribute a significant discovery to an individual person at a particular point in time, but philosopher Samantha Copeland argues that these narratives tend to ignore the communal aspect and contributions. Instead of reinforcing the narrative of the scientist as a lone genius, therefore, Copeland suggests that scientific discoveries are more complex and often recognised retrospectively, when its value and impact have become more evident. Moreover, she notes, 'contemporary science is complex, both in terms of theory and tools—how can any one scientist have the necessary knowledge and resources to achieve progress, without the help of others?' (Copeland, 2017, p. 2399). In a letter from 1675, Isaac Newton famously wrote 'If I have seen a little further it is by standing on the shoulders of Giants', suggesting that his contributions to science built on the knowledge of other great scientists. Copeland proposes that the scientific community ought to be better structured in a way that produces more opportunities for making valuable discoveries and maximise their benefits for scientists, science, and society, such as sharing of knowledge. This requires that its members cultivate the skills, wisdom, and preparedness that are needed to recognise and take advantage of those opportunities.

We will now present an historical case of a ground-breaking discovery that can illustrate several of the ideas presented so far: how scientific activity is a joint effort toward a shared goal; how new knowledge relies on a body of existing knowledge; how new observations fill in missing pieces of a puzzle, and; how scientific discoveries can develop from unexpected observations, if such observations are picked up and acknowledged by a prepared scientific community.

An Example from Biochemistry: Discovering the Structure of DNA

The discovery of the three-dimensional double helix structure of DNA is famously attributed to James Watson, Francis Crick, and Maurice Wilkins who in 1962 were awarded the Nobel Prize for the achievement. However, there is general agreement that this was a community discovery, made possible by a considerable amount of knowledge that had been accumulating over decades. The chemical composition of DNA, for instance, was long known, and its discovery dates back to the 1860s. Biochemists Phoebus Levene and Erwin Chargaff had not only successfully isolated the DNA molecule, but also understood that it is composed of a chain of smaller molecules, or *nucleotides*, as Levene named them. Levene correctly inferred that each nucleotide is composed by three parts: a sugar, a phosphate group, and one of four possible nitrogen bases—adenine, cytosine, guanine, or thymine (Pray, 2008). As a heritage from these early studies, there was an established understanding that DNA is made of chains of sugars bound to each other and the nitrogen bases somehow attached to them. However, there was no idea of how such chains are put together in a molecule, and of how many chains make up a molecule. The structure of DNA remained a puzzle.

A key contribution to the final solution was the famous Photograph 51, taken by biochemist Rosalind Franklin, who specialised in X-ray crystallography, and her PhD student Raymond G. Gosling (see Fig. 3.1). This was a relatively new technique, discovered by William and Lawrence Bragg. The technique was used to study the structure of crystals by bombing them with X-rays and look at the way they were refracted. Soon after its discovery, in the late 1930s, X-ray crystallography was first used to understand the structure of DNA by William Astbury's group at University of Leeds. Compared to the pictures produced by Franklin in the 1950s, Astbury's pictures were blurry, but allowed them to discover the distance between the nitrogen basis in the DNA molecule. This was another important piece of information for the final discovery of the molecular structure.

One of the greatest discoveries that Franklin made was that there are two forms of DNA: a dehydrated and a hydrated, which produced different X-ray patterns. This discovery, that DNA could be in a hydrated form, gave Franklin the clear understanding that the sugar chains (the 'backbone' of the DNA) had to be on the outside of the molecule and not close to each other (otherwise the water molecules could not be contained within it). When Watson and Crick proposed a model of DNA with the sugar-phosphate backbone on the inside and the bases on the outside, Franklin could immediately reject that based on her crystallography experiments. Franklin's Photo 51 was interpreted by Watson, Crick, and Wilkins as a clear image of a helix seen from the side. From more of Franklin's data, Crick was able to deduce that the chains in the DNA molecule look the same upside-down and must therefore run in opposite directions. The final clue was provided by a 1949 experiment by Erwin Chargaff, determining that the number of adenines and thymines was always equal, and so were the numbers of cytosine and guanine. Watson and Crick realised that

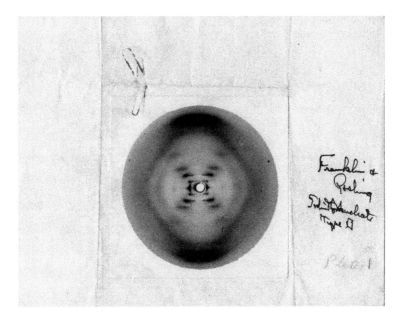

Fig. 3.1 Photograph 51. © 2015 Oregon State University Libraries

adenine must always bond to thymine, and cytosine to guanine, forming the steps of helix-shaped ladder where the sides are formed by two sugar chains (see Fig. 3.2).

From this story, we can see that the narrative of the famous duo and their groundbreaking discovery could easily be retold as the story of the joint success of a prepared community of scientists working together toward a common goal. How would the story be different if the scientists involved did not share the same goal, or even disagreed about whether it was a goal worth pursuing? What if they couldn't agree about which methods or tools were appropriate for the task, or which theoretical framework to assume? This is the reality that scientists often face when a complex problem forces them to collaborate across disciplinary boundaries. When scientific frameworks and research traditions collide, there will be no common set of premises that everyone accepts. In recent years, a number of so-called wicked problems have emerged that require that science moves beyond the disciplinary and paradigm-specific science that Kuhn described as normal science.

Post Normal Science: When Science Requires the Whole Community

Recall the empiricist ideal of science as objective, neutral, and value-free, promoted by Bacon and the logical empiricists. This would allow a strict separation between scientific facts, on the one hand, and societal values, on the other. For a long time,

Fig. 3.2 DNA structure as the double helix (Wikimedia Commons)

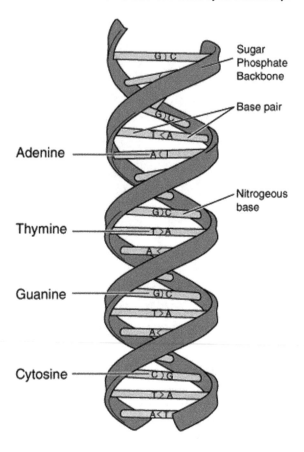

Adenine

Thymine

Guanine

Cytosine

Sugar Phosphate Backbone

Base pair

Nitrogeous base

scientific advances and technological innovations were seen as largely beneficial for humankind, improving life standard and increasing life expectancy. Nowadays, however, we see also the negative impacts of many of those innovations. Recently, the symptoms of an epochal shift to the Anthropocene have made the question of environmental risk a central concern. Industrial agriculture, oil drilling, and new technologies are widely reputed to have a range of unwanted effects. For instance, while the expansion of large-scale industrial agriculture has enabled food security, modern techniques and the use of pesticides are also linked to serious biophysical changes and human health problems. Oil- and petrochemical production has led to the dumping of millions of barrels of chemicals, drilling fluids, and formation water into our seas, rivers, and forests.

When a problem reaches across socio-political, environmental, and economic issues, such as climate change, pandemics, and natural hazards, they cannot be solved by the strategies used in normal science. In their paper, 'Science for the post-normal age', Silvio Funtowicz and Jerome Ravetz argue that we have entered a new era for science. Typical for the era of post normal science, they say, is that *facts are uncertain, values are in dispute, stakes are high, and decisions are urgent*. There will no longer be expert consensus about how to frame a problem, how to solve it, or whether it is the right problem to target. While the scientists within a Kuhnian paradigm would be working toward a common goal, solving the problems and puzzles set by the paradigm, we are now faced with scientific controversies that include conflicting facts, needs, and values.

An example might help illustrate the difference between normal and post normal science. Many scientists are now working to eradicate hunger by increasing food production in the safest possible way. Although such challenges require a multi-disciplinary effort, including the development of agricultural biotechnology, irrigation techniques, fertilisation methods, precise weather forecast, and so on, the scientists work toward the same goal. In the post-normal era, however, even the framing of the problem is challenged. One might disagree over whether assuring an adequate food consumption for the global population is a matter of increasing crop yield, or a matter of governance, global distribution of resources, skill development, subsidies for small farming, or preservation of soil resources and ecosystem services.

In post normal science, Funtowicz and Ravetz explain, we need a radical form of trans-disciplinarity that the applied and disciplinary sciences are not prepared for yet. Science can no longer be seen as separated from society and politics, and any idea of community must therefore include all stakeholders of science. 'Expertise' cannot be defined narrowly, reserved for scientists or consultants, but must include knowledge and expertise from an 'extended peer-community'. Fern Wickson, who works in ethics, governance of new technology, and nature conservation, emphasises the importance of an extended peer-community in risk assessment on new technologies. In her own experience from impact assessment of new technologies, she finds that scientific innovations are dependent on relational networks to fulfil their purpose and function, and that information from these networks ought to be considered and included when evaluating risk. In 'The troubled relationship between GMOs and beekeeping', Rosa Binimelis and Wickson present a case from their work with assessing the consequences of intensive cultivations of genetically modified soybean in Uruguay. Here, beekeepers provided valuable information of how bees are massively suffering from extensive genetically modified cultivations, not because of the technology itself, but because of the loss of flowers caused by the technology. In lab toxicological studies, genetically modified proteins contained in soybeans are found to be innocuous for bees. In agricultural practice, however, they cause a loss of biodiversity that damages the pollinators. Knowledge from local producers, as well as local politicians and interest groups, thus seems vital for uncovering a range of unexpected impacts when applying a general measure locally.

Chapter Summary

In this chapter we have looked at various ways in which science could be seen as a social activity. Knowledge and expertise should not be narrowly defined as belonging to individual scientists, and we have seen here that many of Kuhn's opponents agree with him on this point. We also saw how some philosophers go further than Kuhn, arguing that knowledge production must involve the wider community. Following Funtowicz and Ravetz, scientists should value the knowledge and perspectives from practitioners, interest groups, local community, indigenous population, and anyone affected by the impact of research. This also means that the general, theoretical knowledge acquired in specific academic curricula is insufficient to carry on the process of scientific investigation in a post-normal context. This is because it lacks the applied awareness, the contextual knowledge, and the informal and unofficial evidence that can be brough to the table by those who are directly influenced by the result of the scientific process. While Kuhn saw normal science as defined by the scientific community, post-normal science should be defined by the whole community. If we accept that science cannot be treated as isolated from society, then we ought to critically consider how power structures and power imbalances in society are reflected in science and technology. In the next chapter, we will do exactly this, and we will find out why Harding famously claimed that *science is politics by other means*.

Further Introductory Reading

To read more about science as a social activity, see Helen Longino's (1990) book *Science as Social Knowledge* and Thomas Kuhn's (1962) classic text *The Structure of Scientific Revolutions*. If you prefer to read an article rather than a book, we recommend Susan Haack's (1998) text 'Science as social? - Yes and no', and Samantha Copeland's (2017) paper 'On serendipity in science: Discovery at the intersection of chance and wisdom'. For an introduction to the idea of post-normal science, the best source is Silvio Funtowicz and Jeremy Ravetz's (1993) original article 'Science for the post-normal age'.

Further Advanced Reading

For a better idea of the theory of competing research programs as an alternative to Kuhn's incommensurable paradigms, see Imre Lakatos' (1978) *The Methodology of Scientific Research Programmes*. If you are interested in the idea that scientific evidence works like a crossword puzzle, see Susan Haack's (2002) paper 'Clues to the puzzle of scientific evidence'.

Free Internet Resources

There is a lecture by Helen Longino (2021) on YouTube, 'Critical contextual empiricism, diversity and inclusiveness' by Weizsäcker-Zentrum Universität Tübingen, where she explains her idea of contextual empiricism in more detail. *Nature Education* has a nice article of the case of jumping genes, 'Barbara McClintock and the discovery of jumping genes (transposons)', written by Leslie Pray and Kira Zhaurova (2012). The case discussed by Rosa Binimelis and Fern Wickson (2019) on GMO and beekeeping, 'The troubled relationship between GMOs and beekeeping', is openly available in *Agroecology and Sustainable Food Systems*.

Study Questions

1. What do you think about Kuhn's idea that science happens within a paradigm?
2. What did Kuhn mean by normal science? Do you think this an accurate description of how science is actually done? Or how it *should* be done?
3. Kuhn compared a paradigm shift with a gestalt switch, where one is looking at the same thing but suddenly with a completely new perspective, like in the duck-rabbit drawing in Chapter 2. Why do you think Kuhn made this comparison?
4. Why do you think some people worry about Kuhn's idea that science is socially defined?
5. What were some responses to Kuhn? Which of them do you find most plausible, and why?
6. How do you understand Funtowicz and Ravetz's idea of post-normal science?
7. Can you think of any societal challenges that would require a post-normal science approach? Explain.
8. How would you answer Haack's question *science as social: yes or no*?
9. What are some advantages of interdisciplinary and transdisciplinary science? Do you see any potential challenges?

Sample Essay Questions

1. Present and discuss Kuhn's idea that science happens within a social framework, or paradigm. Define some central notions (e.g., scientific paradigm, normal science, ad hoc hypotheses, gestalt switch, incommensurability) and explain what role they play in Kuhn's theory of science. Include some of the responses to Kuhn. What is your own response?
2. What is the role of the community for science and knowledge production? Referring to some of the philosophical theories presented in this chapter, discuss the idea that scientific progress cannot be made by individual researchers or even isolated scientific communities.

3. Compare and discuss disciplinary, interdisciplinary, and transdisciplinary approaches to science. Do all types of problems require that scientists work across disciplinary boundaries? What do you see as good arguments for each of these scientific approaches, and what do you see as potential challenges?

References

Binimelis, R., & Wickson, F. (2019). The troubled relationship between GMOs and beekeeping: An exploration of socioeconomic impacts in Spain and Uruguay. *Agroecology and Sustainable Food Systems, 43*, 546–578.

Bourque, G., Burns, K. H., Gehring, M., et al. (2018). Ten things you should know about transposable elements. *Genome Biology, 19*, 199.

Buehler, J. (2021, September 1). The complex truth about "junk DNA". *Quanta Magazine.* https://www.quantamagazine.org/the-complex-truth-about-junk-dna-20210901/#

Copeland, S. (2017). On serendipity in science: Discovery at the intersection of chance and wisdom. *Synthese, 196*, 2385–2406.

Funtowicz, S., & Ravetz, J. (1993). Science for the post-normal age. *Futures, 25*, 739–755.

Haack, S. (1996). Science as social? Yes and no. In L. H. Nelson & J. Nelson (Eds.), *Feminism, science, and the philosophy of science*, Synthese Library (Studies in epistemology, logic, methodology, and philosophy of science, vol. 256). Springer.

Haack, S. (2002). The same, only different. *Journal of Aesthetic Education, 36*, 34–39.

Kuhn, T. (1962). *The structure of scientific revolutions.* University of Chicago Press.

Lakatos, I. (1978). *The methodology of scientific research programmes* (Philosophical papers: Volume 1, J. Worrall & G. Currie, Eds.). Cambridge University Press.

Longino, H. (1990). *Science as social knowledge: Values and objectivity in scientific inquiry.* Princeton University Press.

Longino, H. (Online Lecture, 2021). *Critical contextual empiricism, diversity and inclusiveness.* Weizsäcker-Zentrum Universität Tübingen. YouTube. https://www.youtube.com/watch?v=XysYymrh7IE

Pray, L., & Zhaurova, K. (2008). Barbara McClintock and the discovery of jumping genes (Transposons). *Nature Education, 1*, 169. https://www.nature.com/scitable/topicpage/barbara-mcclintock-and-the-discovery-of-jumping-34083/

Pray, L. A. (2008). Discovery of DNA structure and function: Watson and Crick. *Nature Education, 1*, 100.

Ravindran, S. (2012). Barbara McClintock and the discovery of jumping genes. *Proceedings of the National Academy of Sciences, 109*, 20198–20199.

Sulston, J., & Ferry, G. (2002). *The common thread: A story of science, politics, ethics and the human genome.* Joseph Henry Press.

Chapter 4
Is Science Defined by Power?

The Power of Science (and Those Paying for It)

So far in part I, we have seen that philosophers of science disagree over what counts as scientific knowledge, what is the best scientific method, and how or whether science makes progress. There seem to be no commonly accepted ways to guarantee objective and true scientific results, and philosophers have questioned whether science can ever be objective or independent of the beliefs shared by members of the scientific community. As we saw in the previous chapter, Kuhn describes science as a socially influenced system. Even many of his opponents seem to agree with Kuhn that what counts as scientific knowledge and results will depend on some sort of consensus within the scientific community about what counts as relevant theories, methods, and results. If so, then who gets to set the premises within the scientific community? Is it a purely democratic matter, to be decided by a majority? Or could there be a powerful minority of scientists that gest to dominate entire professions, or worse, all of science? In what way do the interests of the powerful influence the scientific agenda and focus? And to what extent is science influenced by external interests, including by those who pay for research?

In this chapter we present some concerns raised by philosophers of science about the objectivity, democracy, independence, and representation of research. We will see that feminist philosophers of science have argued that social and political power imbalances influence how the costs, benefits, and relevance of science and technology are unevenly distributed among members of the global society. If science is shaped by a powerful and privileged minority, then what are its biases and blind spots? These are some of the issues that will be addressed here.

One philosopher of science concerned with science and power is Paul Feyerabend. In his provocative lecture, 'How to defend society against science', Feyerabend argues that science holds a powerful role in society to such an extent that it has become our new religion. Unlike religion, however, which is typically full of diverse theories and controversies, science is a belief system that mainly involves likeminded

© The Author(s), under exclusive license to Springer Nature Switzerland AG 2024
R. L. Anjum and E. Rocca, *Philosophy of Science*, Palgrave Philosophy Today,
https://doi.org/10.1007/978-3-031-56049-1_4

believers: the community of scientists. While all ideologies are subject to opposing views, science holds a privileged status in society, where it is excluded from criticism. Alternatives to science are systematically dismissed only and precisely because they are not scientific. This, he thinks, is not a role that science deserves or should even want to hold. How does he defend this claim?

History, Feyerabend says, tells us that science saved us from the authority and superstition of religion and brought enlightenment instead. Where religion is seen as dogmatic and unquestioned, science is presented as a systematic, self-critical, and objective search for truth. Surprisingly, perhaps, Feyerabend rejects any such clear-cut distinction between science and religion. Today, he notes, scientific facts are taught in schools in the same way that religious facts used to be taught only a century ago: as undeniable truths. University education is even worse, he thinks, and has become more about theoretical indoctrination than critical reflection. The holy book in the teacher's hand is replaced by the science book, taught with the same conviction and dogma and as immune to criticism. To silence opposition, all one has to say is that it is *unscientific*. Feyerabend sees this as a sign of the great power and authority that science has gained in society, which he thinks is not a healthy development. 'Science has now become as oppressive as the ideologies it had once to fight' (Feyerabend, 1975a).

Is this comparison between science and religion too strong, he asks? After all, people used to be beheaded for religious blasphemy back in the days, something that could not happen in our civilised society. This might be so, Feyerabend admits, but although no one gets killed for opposing science, heretics of science still suffer the most severe punishment available in society today. What does he mean by this? Who gets punished for scientific heresy, and who is punishing them? What could possibly be the motivation behind such punishment?

Let's pause for a moment here and consider the important role that science holds in our modern society. We all rely on science to help us solve the biggest challenges facing our times, such as global warming, loss of biodiversity, pollution of air and water, pandemics, broken food systems, and poverty. These are problems that affect the global community and require urgent attention. As stakeholders of science, we all share an interest in the outcomes of science. When someone tries to push science on the side-line and replace it with superstition or dogma, we as a community must speak up against it and act. Medicine should not be taken over by quacks, evolutionary theory should not be replaced by creationism in school education, and no enlightened society can accept that ideology distorts scientific facts. While this might seem entirely obvious and uncontroversial, it is worth considering to what extent science itself is influenced by internal and external powers and interests. If, as Feyerabend claims, science has a privileged and powerful position in society, then who holds the power over science? We are all stakeholders of science, but it's not obvious that everyone's interests in science counts equally. One reason for this is the high stakes of modern science, which is not only truth, not only religion, but also money. Let us explain.

Research requires funding, and today most universities and research institutions rely heavily on the income from commissioned research by public and private sector.

That research is commissioned means that the funding comes from external sources and that the client decides which problem to address, while also retaining rights to use the results. Supported by international research ethics guidelines and legislations, the scientific community holds the responsibility of making sure that the integrity of research is not compromised by external interests and power. Values in science include academic freedom and independence, trustworthiness, accountability, and openness. With the commercialisation of research, however, these values are threatened, especially if there are strong interests involved in what results are produce and who should have access to them.

One important interest in research and innovation lies in the patent laws. Unlike resources that exist naturally, such as food and plants, all artificially produced products can be patented and made into big business. The company that has the patent will then hold the intellectual and financial rights to control production and sales. During the COVID-19 pandemic, vaccine patents were highly disputed because poorer countries could not afford to buy all the vaccines they needed, and the patent laws prohibited them from making their own vaccines faster and cheaper. On the other hand, many pharmaceutical companies and rich nations argued that if it weren't for the patents, there wouldn't be the same financial incentive to develop new vaccines in times of crisis.

Today, an increasing concern is raised about a controversial patent practice that its critics call 'biological piracy', 'biopiracy', or 'scientific colonialism': the act of patenting native plants or other biological material. The colonialism lies in the fact that these plants are normally found in the resource rich global south, while the patents tend to be taken out in technologically advanced countries in the global north. This has happened with a number of plants traditionally used for food or medicine, and examples include aloe vera from Thailand, Darjeeling tea and turmeric from India, rooibos tea from South-Africa, enola beans from Mexico, and the teff grain from Ethiopia. Biopiracy is made possible by a loophole in the patent law, allowing patenting of slightly processed versions of the native plant. The highly nutritious and gluten-free teff grain had been used by Ethiopians for over 3000 years when a Dutch company, Health and Performance Food International, was granted patent by the European Patent Office in 2006. Their patented invention was on 'the processing of teff flour'. After years of legal battle, The Hague court decided in 2019 that the patent on teff should be withdrawn in Europe (Wilhelm, 2020).

Commercial interests in science seems to be the main motivation for a significant proportion of research. An increasing number of scientific studies are now paid by external funders with a direct or indirect financial interests in the outcomes of research. Biological research might be funded by biotech companies that hold patents of genetically modified plants. Geological research might be funded by the oil- and gas industry and the governments that own them. Medical research might be funded by the pharmaceutical industry that holds patents of vaccines and medicines. Aquaculture research might be funded by fish farming industry. Problems arise when scientists let the expectations of clients and investors compromise the research process, for instance to secure continued financial support or career advancement, or to avoid conflict.

Scientific disputes and lawsuits involving industry are not uncommon. One example is the involvement of Exxon, one of the world's largest oil- and gas companies, in climate science. In the 1970s, when the scientific evidence started to show the negative impacts of carbon dioxide emissions released from burning fossil fuels, Exxon contributed to that research. Since then, however, the company has played an active role in discrediting existing climate research by focusing on expert disagreement and uncertainty (McNeish et al., 2015). This strategy is best known from the tobacco industry's involvement in the scientific dispute over whether smoking causes lung cancer. To discredit the available medical evidence, the tobacco companies were advised by a consultant agency to counter it with its own scientific research. In 1953, the Tobacco Industry Research Committee was established, which later was renamed Council for Tobacco Research. Instead of studying the link between smoking and lung cancer, however, the council funded research that explored other potential causes of cancer: heredity, infection, nutrition, hormones, nervous strain or tension, and environmental factors. Up until 1998, the committee carried out 300 million USD worth of research paid for by the tobacco industry. In 1988, the Center for Indoor Air Research was established by three major tobacco companies, to counter the scientific evidence of health hazards from second-hand smoking. The research centre funded over 240 studies on air pollution, many of which focused on environmental toxins other than passive smoking, and on confounding factors such as genetic pre-disposition, diet, and stress (Muggli et al., 2001). The aim was to show that, compared to other environmental stressors, second-hand smoking was an insignificant factor. The Council for Tobacco Research and the Centre for Indoor Air Research were dissolved in 1998, as part of the Tobacco Master Settlement Agreement of a 40-state lawsuit against the industry to cover tobacco-related healthcare costs. This is not a unique case, although it is the most famous example of how science can in principle comply with established methods and standards but still be used to mislead the public (Schick & Glantz, 2007).

Returning now to Feyerabend's passing claim, that heretics of science receive the strictest form of punishment that our civilization allows. Who gets punished in science today, and what punishment can they receive? When corporations are paying for researchers, equipment, and even university buildings, they hold a powerful position that can be exploited. If an institution is in danger of losing research funding because of unwanted results, it's easier to get rid of individual scientists and their data than to publish the results and face the financial consequences. If one follows the money and sees who benefits from the scientific results, one might find that there are heavy interests that sway the scientific debate on what is safe or harmful.

In what has become known as the Pusztai affair, the scientific dispute were over what conclusions to draw from a pioneering study assessing the safety of genetically modified potatoes when fed to rats. In a television interview, lead scientist Árpád János Pusztai was invited to discuss the findings of his team's research (Ewen & Pusztai, 1999). The institute had approved and encouraged participation in the interview, and according to Pusztai, the goal was to attract commercial funding after the project ended. In the interview, Pusztai mentioned his concerns about the findings of stunted growth and repression of the rats' immune system. He also questioned the appropriateness of the research methods for concluding about the safety of genetically modified foods and remarked that he himself would not choose to eat it. The interview led to a media storm and the institute immediately received several calls, including from the Royal Society and the Prime Minister's Office. After the interview, Pusztai was faced with charges of scientific misconduct from his own research institute, got suspended from his position, had his data confiscated and team removed, and he was banned from speaking publicly about his findings. His annual contract was not renewed.

Irrespectively of the details in this particular dispute, the Pusztai affair illustrates Feyerabend's point that stakes are high in science and that the scientific community has powerful sanctioning methods for silencing researchers. We will return to Feyerabend's suggestions for how to liberate science from dogma later in this chapter. Before this, we will look at another and perhaps more abstract power figure in science: the reference man. With the reference man in charge, science has been criticised for excluding and marginalising most of the world's population in their data, leaving a significant knowledge gap and biased results. A worry is that science represents an already powerful and privileged minority of the global society.

The Power of the 'Reference Man' in Science

In March 2019 NASA had to cancel their plans for a first all-female spacewalk for reasons of safety concerns, making international headlines. The reason was that NASA only had one spacesuit in size medium. Since cosmonaut Valentina Tereshkova had become the first woman in space in 1963, only 11% of all space travellers have been women. Despite the gender imbalance in the history of physical space exploration, both astronauts of the planned all-female spacewalk had been part of the NASA 2013 class, where half the class were women. The lack of spacesuits in their size might therefore seem surprising, while in fact this failure to accommodate women's bodies reflects a global trend within science and technology to rely on a male default. In medicine, for instance, the 'reference man' is a 20–30 year old male who weighs 70 kg and is 170 cm tall. This model was initially developed for estimating safe levels of radiation doses, but has been widely used also in pharmacology, nutrition, and toxicology to estimate dose-response relationships. However, we know that any reaction to a given exposure will vary, depending on the physiological properties of the exposed individual. From the same exposure, a healthy adult might be less vulnerable to harm than a malnourished child, or than an older person with a chronic condition. This is why, when generating the scientific estimates of harm and safety from some assumed norm or average model, one risks getting it wrong for anyone who deviates from that standard.

In her book, *Invisible Women: Exposing Data Bias in a World Designed for Men*, Caroline Criado Perez explains how women pay tremendous costs of the male data bias in science and technology, in time, money, and often with their health and lives. When stab vests, crash test dummies, protective face masks, and CPR dummies are designed to fit the reference man, it puts many people at risk. For instance, several studies have revealed that women have a much higher risk of moderate and serious injury from car crashes than men from the same car crash, and even a higher risk of death. Since the reference man is commonly used to estimate safe levels of chemical exposures, these estimations won't have universal application. Perez also notes how much of medical research is using data from men, male animals, and even male cells. Is this because one believes that one size fits all? It seems not. One argument for collecting data primarily from males is that women's hormones and immune system are more active and tend to interfere with the intervention and give different scientific results. According to Perez, there is only one rational response to this: that we need data also from women in order to get scientific results that are applicable to women. Another reason why many women have been excluded from medical trials is to avoid another undesired consequence similar to the thalidomide disaster, where the foetus got harmed during early pregnancy (see ch. 2). To avoid the risk of unpredictable damage to the foetus, the FDA decided in 1977 to exclude 'women of childbearing potential' from pre-marketing clinical trials of medicines, a decision that was reversed in 1997. In that period, there were no exceptions to this, so women of fertile age were excluded even if they were single, used contraception, or their partners had been vasectomised.

Data bias in science is a problem that not only affects women. The original reference man, as defined by the International Commission on Radiological Protection (ICRP), is Caucasian with a Western European or North American habitat and custom, who lives in a climate with an average temperature of 10–20 degrees Celsius (ICRP, 1975, p. 4). In a 500 pages report, all of reference man's characteristics are specified in full detail, including the volume, dimension, and weight of all tissues and organs. The reference man served a specific purpose: as a scientific standard for estimations of doses, comparisons of results, and safety recommendations. The authors specify this in the report:

> Only a very few individuals of any population will have characteristics which approximate closely those of the reference man, however he is defined. The importance of the reference man concept is that his characteristics are defined rather precisely, and thus if adjustments for individual differences are to be made, there is a known basis for the dose estimation procedure and for the estimation of the adjustment factor needed for a specified type of individual. (ICRP, 1975, p. 4)

Although the reference man was not supposed to represent everyone, his influence in medical standardisation seems evident. One example is how the dosage of blood donations is determined. The original reference man from 1974 has 5,3 litres of blood, while the reference woman has 3,9 litres (ICRP, 1975, p. 33). However, blood volume also varies with hight and weight. Scandinavians, for instance, tend to be taller than Indians, so will tend to have more blood. This is reflected in national guidelines: in India, one must weigh over 45 kg to donate blood, while the minimum weight in Norway is 50 kg. The default blood donation volume, however, is usually the same (450 ml to 500 ml) and is set to match the standard anticoagulant mix proportions added to the donated blood. An alternative approach would be simply to adjust the standard to the relevant population or offer two standards, as in India where one donates 350 ml or 450 ml. That way, more people can make donations without facing health risks.

When vital physiological and anatomical knowledge is generated from the reference man, it affects the relevance of research and technology for much more than half of the world's population. One example is how well (or poorly) face recognition technology works for different population groups. A study of the accuracy of AI powered gender classification products, carried out by the Gender Shades research project, found significant gaps in error rates between men and women, with significantly worse results for young women of colour (Buolamwini & Gebru, 2018). A reason for this, they note, is that some of the most commonly used data sets for training software algorithms are predominantly white and male, such as the Labeled Faces in the Wild database (Han & Jain, 2014). Just like a gender bias in the data will tend to result in algorithmic gender discrimination, a racial bias will impact the accuracy of predictions for certain ethnic groups. When this type of technology is used extensively, for law enforcement surveillance, airport passenger screening, or for controlling access to buildings and facilities, it can have serious effects on people's lives. In response to this, many researchers are now asking that more attention is given to how data bias in artificial intelligence and machine learning is affecting already marginalised groups. In *Data Feminism*, Catherine D'Ignazio and Lauren F. Klein emphasise the need of data science to examine and challenge existing power structures as part of their work. They ask: *Data science by whom? Data science for whom? Data science with whose interests and goals in mind?* and propose several strategies for the field to move forward, to counter data bias and injustice.

Another relevant example of data bias in science is found in psychology and behavioural science. This is significant, since these disciplines have major influence on any technology or tools designed to predict and influence people's preferences, decisions, and actions. For decades, however, research in both these fields has been dominated by data collected from a very small segment of the global society: so-called WEIRD populations. WEIRD is an acronym first used in 2010 by Joseph Henrich, Steven Heine, and Ara Norenzayan to express the fact that most people recruited to behavioural science studies are graduate students from Western, Educated, Industrialised, Rich, and Democratic backgrounds. In their research, they found that this group only represents about 12% of the global population. This means that any conclusions based on those data cannot be generalised, and that theories in psychology and behavioural science might have limited relevance and predictive power for most people and cultures. Although WEIRD populations are a minority group, it is a group that holds a strong position in both science and society.

When gender, racial, and other forms of discrimination in data result in disproportionally wrong estimates for socially marginalised groups, it raises concerns about the assumed objectivity of science and technology. Can science ever be neutral or bias-free? Which groups are represented in research? Who benefits from scientific knowledge and technological innovations? If social inequalities manifest themselves in technologies that were intended to reduce error caused by human bias, these technologies will unintentionally reinforce those inequalities. Many philosophers of science worry about data bias and lack of representation in research. One obvious danger is that a vast majority of people risks being marginalised by research that is done by and for a privileged minority. In the previous chapter, we discussed the view that science is defined partly or wholly by the scientific community. If it is the community of scientists that decides which theories, authorities, methods, concepts, tools, and even knowledge gaps are worth our attention and resources, it seems at least relevant to ask which groups are represented in those communities. This is a question that has been raised by feminist philosophers of science.

Whose Science? Whose Knowledge?

In philosophy and beyond, we now see a growing awareness of how social power imbalances can influence scientific knowledge. Feminist philosophers of science have challenged the logical positivist idea that science could ever be neutral or value-free. For instance, among all questions and problems, only some will count as scientifically significant and worthy of pursuing, which is primarily a question of interest, priority, and value. Financing of research is another important issue. Since funding is limited, one needs to consider carefully where the money is best spent. Who gets to make these decisions? What is their social-economic status? And whose interests do they take into account when doing so? These matters cannot easily be settled without making some value judgements.

Recall how the logical positivist ideal of science was inspired by empiricists, such as Francis Bacon (1620) and his new method for a purely data-based science. While Bacon had good reasons for wanting a neutral and value-free science, namely, to liberate science from dogma and religion, he was acutely aware of how science is influenced by personal, cultural, conceptual, philosophical, and other human biases. Bacon's solution was for the scientists to first become aware of their biases, and then to work systematically to avoid them while doing research. But even if it were possible to free oneself from personal biases, what about all the assumptions, perspectives, and blind spots that one shares with the whole discipline, or even the broader scientific community? Many disciplines share an empiricist bias, for instance. It is then difficult to even detect empiricism as an assumption one makes, rather than as an integral part of science itself. For a long time, there was a shared disciplinary bias toward mind-body dualism in medicine that no one thought to challenge. Would these shared biases even be seen as flaws or shortcomings within a Kuhnian paradigm, or instead as necessary premises for doing normal science?

Some philosophers argue that, instead of pretending that the personal, cultural, and paradigmatic influences aren't there, we need an increased awareness of how science mirrors the biases and interests of those who dominate science. In *Whose Science, Whose Knowledge. Thinking From Women's Lives*, Sandra Harding argues that it is a democratic problem when science and research primarily represent perspectives and interests of dominant groups in society. If those who dominate society also dominate science, and science has a prominent role in society, then lack of representation is a serious issue that needs to be addressed. She promotes what she calls 'strong objectivity' in science, which is neither a positivist nor a relativist notion. Scientific claims should not be thought of as 'weakly objective': neutral, impartial, or dispassionate. This weak, positivist notion of objectivity, she argues, misses the point that the social values and interests that are excluded will always be judged by the members of the scientific community. On the other hand, it does not follow that scientific claims are subjective. A strong form of objectivity can only be gained when we welcome perspectives that help identify and challenge assumptions and agendas that are tacitly accepted by the scientific community. Scientific research should include the systematic examination of background beliefs (Harding, 1991, p. 149). In order to do this, one must start from the lives of those whose perspectives and values have been excluded: the marginalised groups. This idea is typical for a standpoint theory of knowledge.

In Chapter 1, we mentioned briefly how standpoint theory sees all knowledge as influenced by one's socio-political context. The Marxist inspiration for standpoint theory is that if we want to understand a system, such as capitalism, we cannot investigate it only from the perspective of those who benefit from it. Members of the working class will experience a different side of capitalism precisely because they are marginalised by it. This is why, Marx says, groups that have no interest in the continuation of a system will have an *epistemic advantage* over the privileged classes in understanding that system. Basically, this means that if we want to fill our knowledge gaps in science, we should include the perspectives and insights of those who bear the costs, not only those who benefit from science. This is the strong objectivity that Harding promotes, and according to many feminist philosophers of science, it is the closest one can get to any form of objectivity in science. We see, then, why feminist philosophy of science often sides with the perspectivist theory of knowledge, denying that there could ever be *a view from nowhere* that the scientist could take. Donna Haraway calls it 'the god trick' when we attempt to step outside of our own perspective and take on a god's perspective. This is an illusion, she says. 'The moral is simple: only partial perspective promises objective vision' (Haraway, 1988, p. 583). Like Harding, Haraway is critical of the way that relativism has been seen as the only alternative to a 'god trick' form of objectivity.

> Relativism is a way of being nowhere while claiming to be everywhere equally. The 'equality' of positioning is a denial of responsibility and critical inquiry. Relativism is the perfect mirror twin of totalization in the ideologies of objectivity; both deny the stakes in location, embodiment, and partial perspective; both make it impossible to see well. Relativism and totalization are both 'god tricks' promising vision from everywhere and nowhere equally and fully, common myths in rhetorics surrounding Science. But it is precisely in the politics and epistemology of partial perspectives that the possibility of sustained, rational, objective inquiry rests. (Haraway, 1988, p. 584)

Standpoint theory comes in many versions, and some of these are more empiricist than perspectivist. According to empiricism, our aim should be to get as complete a data set as we can possibly achieve. The more facts we get, the more accurate our theories, explanations, and predictions will be. Faced with the problem of how to fill in knowledge-gaps relevant for marginalised populations, the empiricist response will therefore be that we need better and more representative data. For this, a plurality of perspectives might help us get a more complete picture. A similar idea was promoted by Hannah Arendt, a highly influential political thinker. Plurality, she says, is a fact of 'the human condition'. We are all different, with our unique experiences, viewpoints, and stories. What makes us human is when we engage openly and actively with others and expose ourselves to different perspectives. Plurality is, according to Arendt, an important source of critical thinking and good judgements. In *The Origins of Totalitarianism*, she explains how plurality differs from ideology, which is typical for totalitarian systems where one is supposed to accept and promote a single, fixed view (Arendt, 1951).

Note, however, that pluralism is not the same as relativism. To say that we must see the world though our own lenses, and that our experiences are situated, does not necessarily mean that there is nothing real to see or experience. For Arendt, critical thinking must be based on *factual truth*, a concept that relies on plural perspectives. Factual truth, she explains, 'is always related to other people: it concerns events and circumstances in which many are involved; it is established by witnesses and depends upon testimony; it exists only to the extent that it is spoken about, even if it occurs in the domain of privacy. It is political by nature' (Arendt, 1967, p. 300). The more perspectives we hear and learn from, the closer we get to the truth. Born a Jew in Nazi Germany, Arendt's witnessed first-hand the way in which Jews and their stories got erased from public awareness through systematic lies by the Nazi state. She also experienced life as a stateless refugee and saw how basic rights are restricted to the rights given by a political government when recognised as a lawful citizen.

What does this have to do with science? Historically, quite a lot. Today we know how thousands of Jews and other prisoners in the Nazi concentration camps were harmed, tortured, disfigured, traumatised, and killed in 'scientific' experiments between 1939 and 1945, too horrendous to describe here. When dehumanised and stripped of their human worth, they were no longer protected by basic human or legal rights. We also know that unethical experimentation on humans is historically carried out on people from vulnerable and marginalised groups, including low-income communities, indigenous populations, prisoners, and orphans. One example is the infamous Tuskegee study which was aimed at studying the natural evolution of untreated syphilis in male African Americans, and was carried out over 40 years by the US Public Health Service (USPHS) up until 1972. The participants did not give informed consent and thought they received treatment for 'bad blood', but in fact they received no treatment at all, even after a cure for syphilis was found.

As noted by feminist philosophers of science, the costs and benefits of science are not equally distributed in the population, but often reflect the socio-political power imbalances ingrained in the global community, even today. Technological innovations toward the green shift have provided us with electric cars, solar panels, and cleaner air, but the costs on health and environment of extraction, production, or waste, are typically paid by poor communities in the global south. Harding urges that we make radical changes to ensure that science benefits everyone, not just the rich and powerful. We need science and technology that are for people of all classes, races, and cultures, 'not primarily for the white, Western, and economically advantaged "male men" toward whom the benefit from the sciences has disproportionately tended to flow' (Harding, 1991, p. 5). For this, she says, we must also look at the diverse and conflicting interests in science, in and among the marginalised populations.

Women's interests are not homogenous, Harding notes, so cannot be treated as they are all the same. It matters, for instance, whether one belongs to a certain class, race, religion, sexual orientation, disability, and so on, which are all common sources of discrimination. When someone belongs to more than one marginalised group or category, they risk being excluded from any interest group that focuses on a single type of discrimination. In 'Demarginalizing the intersection of race and sex', Kimberlé Crenshaw introduced the term 'intersectional' to describe the situation of being marginalised in more than one way. The intersectional experience of black women, she notes, is greater than the sum of racism and sexism. 'When feminist theory attempts to describe women's experiences through analyzing patriarchy, sexuality, or separate spheres ideology, it often overlooks the role of race. Feminists thus ignore how their own race functions to mitigate some aspects of sexism and, moreover, how it often privileges them over and contributes to the domination of other women. Consequently, feminist theory remains *white*, and its potential to broaden and deepen its analysis by addressing non-privileged women remains unrealized' (Crenshaw, 1989, p. 154). Her notion of intersectionality highlights the complex interconnections between privilege, power, and oppression.

When social and political power imbalances are reflected in science, this is not only a democratic problem, as Harding notes, but a scientific one. Lack of empirical data from most of the global population makes scientific results less reliable and their application limited. This is a methodological concern that feminist philosophers of science have placed on the agenda, but that is not sufficiently acknowledged by the scientific community, or even in contemporary mainstream philosophy of science.

Scientific Controversy and Opposition

Feyerabend was worried about the privileged role science has gained in society, if science is thereby excluded from criticism. He promoted intellectual freedom and argued that modern science inhibits freedom of thought. The best way to prevent science from becoming a dogmatic and oppressive ideology, he thought, is for it to be challenged by opposing views and criticism. Opposing views are not always welcomed in science, however, since the general expectation is that science should deliver truth, not diversity of opinion. But even if this is the case, truth is not a good enough reason to exclude criticism, should we believe Feyerabend. Historically, what counts as truth in science has changed radically with new theories and observations, and it is only by comparing theories that one can decide which is better. We therefore need *a battle of ideas* to advance scientific thinking.

Truth and falsity are not the main concern here, but intellectual freedom and pluralism, ensuring that no theory stands unchallenged. 'Any ideology that makes man question inherited beliefs is an aid to enlightenment. A truth that reigns without checks and balances is a tyrant who must be overthrown, and any falsehood that can aid us in the overthrow of this tyrant is to be welcomed', he says (Feyerabend, 1975a). The aim is not to reach consensus in science, but to counteract dogma. Scientific progress cannot happen if we remove opposition. Feyerabend reminds us that science has often been advanced by people who considered themselves outsiders, and that we need those outsider perspectives to do good science.

Chapter Summary

We have presented some concerns about power imbalance and objectivity of science, and the problem of data bias and data gaps. When the interests and perspectives of marginalised populations and minorities are disregarded also in science and technology, then something needs to change. We discussed the importance of representation, democracy, and plurality of standpoints as ways to remedy the situation and ensure what Harding and others calls *strong* objectivity. Plurality is also an important element of scientific discourse, and we heard how Feyerabend argues that critical voices should be welcomed rather than oppressed. Expert disagreement could then be constructive. In 'Ethics of science for policy in the environmental governance of biotechnology', Fern Wickson and Brian Wynne (2012b, p. 331) advise that debates among experts are conducted with honesty, accuracy, and consistency, and open for discussions that include all stakeholders of science. The alternative to such openness and inclusion is a polarisation of views, where each side digs themselves deeper into their own position and seeks further evidence to support it. This type of expert disagreement is not productive, and it also tends to exclude stakeholders who are less informed or lack the tools to evaluate the disagreement critically. When a scientific issue becomes polarised, one might be left with the option of taking sides depending on who one trusts more. Next, in part II, we offer some philosophical tools for analysing scientific arguments and controversies in a way that makes their implicit premises more explicit and open for critical reflection by all stakeholders of research.

Further Introductory Reading

To learn more about data gaps and data bias in science, we recommend Caroline Criado Perez's (2019) *Invisible Women. Data Bias in a World Designed for Men*, for which she was awarded the Royal Society Science Book prize, and *Data Feminism* by Catherine D'Ignazio and Lauren F. Klein (2020). Both are easy to read and engaging books. For a more philosophical perspective on power and democracy in science, see Sandra Harding's (1991) classic text *Whose Science? Whose Knowledge?* and Donna Haraway's (1988) 'Situated knowledges: The science question in feminism and the privilege of partial perspective'. Another text that is truly worth reading is Paul Feyerabend's (1975a) lecture 'How to defend society against science', where he explains why science should welcome critical voices and plurality of perspectives.

Further Advanced Reading

To dig deeper into the issue of racial and gender bias in science and technology, see Kimberlé Crenshaw's (1989) famous paper 'Demarginalizing the intersection of race and sex', and 'Age, gender and race estimation from unconstrained face images' by Hu Han and Anil Kumar Jain (2014). Joseph Henrich, Steven Heine, and Ara Norenzayan (2010) introduce the WEIRD acronym in 'The weirdest people in the world?'. In *Psychology's WEIRD Problems*, Guilherme Sanches de Oliveira and Edward Baggs (2023) argue that not much has been done in the field to improve the situation and that the problem with lack of representation in study participation is deeply rooted in theory, methods, and institutional structures. If you are interested in Hannah Arendt's notion of pluralism, we recommend *The Origins of Totalitarianism* (1951), a classic text that has gotten a revival in the post-truth era. Her 'Truth and politics' (1967) is also worth a read, as well as her book *The Human Condition* (1958).

Free Internet Resources

Stanford Encyclopedia of Philosophy has an excellent open access article, 'Feminist epistemology and philosophy of science', written by Elizabeth Anderson (2000, 2020). Feyerabend's 1975 lecture 'How to defend society against science' is freely available at different websites. You will also find the full text where Crenshaw (1989) coins the term 'intersectionality'; 'Demarginalizing the intersection of race and sex'. Criado Perez has a podcast that is worth following, called 'Visible women'. To learn more about the controversy over patent laws, see the European initiative 'No patents of seeds', the SDHS research report (2018) 'Status of patenting plants in the global south', and the film 'Seed: The Untold Story' (2016).

Study Questions

1. What do you think about Feyerabend's comparison between science and religion, and his claim that science has taken the place of religion in modern society?
2. Can you think of any areas in research where there could be external pressure to produce the 'right' type of scientific results?
3. What did Feyerabend mean by the battle of ideas, you think? How can it increase democracy in science if it welcomes a plurality of ideas and perspectives?
4. What do you think Harding meant by science being politics by other means?
5. What is your reaction to the problem of gender and racial bias in science and technology discussed in this chapter?

6. Harding argues that science has a democratic problem. If dominant groups in society also dominate science and technology, which again dominates science: How does this influence the way science and technology is developed?
7. In a famous public speech from 1851, Sojourner Truth asks: 'Ain't I a Woman?' As a black women and former slave, she felt excluded from the women's rights discourse. In light of this, discuss Crenshaw's concept of intersectionality.
8. Pregnant women and potentially pregnant women were excluded from medical trials to protect them from risks in the aftermath of the thalidomide scandal. Which other groups in the global community might have been excluded from medical research, do you think?
9. What would you advise to be done to ensure representation and democracy in science?
10. Science and power has been the main topic in this chapter, and one powerful aspect of science is money and funding. How do you think this affects scientific enquiry, intellectual freedom, and public trust in scientific results?

Sample Essay Questions

1. Present and discuss some of the criticism that has been raised against objectivity in science and technology. Explain Harding's notion of strong and weak objectivity and compare this to the idea of plurality, as discussed by Feyerabend, Arendt, and others. Make sure to include your own perspectives.
2. Discuss data gaps and data bias in science in light of the views presented in this chapter. Include a discussion of Crenshaw's term 'intersectionality', and Heinrich and colleagues' notion of WEIRD populations. Finally, what do you think can be done to ensure better representation and social justice in science?
3. Discuss the link between science and money, and which challenges this might represent for scientific enquiry, intellectual freedom, and public trust in scientific results. Refer to the views presented in this chapter and use historical or contemporary examples of scientific controversy.

References

Anderson, E. (2020). Feminist epistemology and philosophy of science. In E. N. Zalta (Ed.), *Stanford encyclopedia of philosophy* (Spring 2020 ed.). https://plato.stanford.edu/archives/spr2020/entries/feminism-epistemology

Arendt, A. (1951). *The origins of totalitarianism.* Harcourt.

Arendt, A. (1967, February 25). Truth and politics. *The New Yorker.* https://www.newyorker.com/magazine/1967/02/25/truth-and-politics

Bacon, F. (1620). *The new organon* (F. H. Anderson, Ed.). Bobbs-Merrill. A free version can be found on the Early Modern Texts webpage. https://www.earlymoderntexts.com/assets/pdfs/bacon1620.pdf

Buolamwini, J., & Gebru, T. (2018). Gender shades: Intersectional accuracy disparities in commercial gender classification. *Proceedings of Machine Learning Research, 81*, 1–15.

Crenshaw, K. (1989). Demarginalizing the intersection of race and sex: A black feminist critique of antidiscrimination doctrine, feminist theory and antiracist politics. *University of Chicago Legal Forum, 140*, 139–167.

Criado Perez, C. (2019). *Invisible women: Data bias in a world designed for men.* Abrams Press.

D'Ignazio, C., & Klein, L. F. (2020). *Data feminism.* The MIT Press.

Ewen, S. W., & Pusztai, A. (1999). Effect of diets containing genetically modified potatoes expressing galanthus nivalis lectin on rat small intestine. *The Lancet, 354*, 1353–1354.

Feyerabend, P. (1975a). How to defend society against science. *Radical philosophy* (pp. 261–271). Stegosaurus Press.

Feyerabend, P. (1975b). *Against method.* New Left Books.

Han, H., & Jain, A. K. (2014). Age, gender and race estimation from unconstrained face images. *Michigan State University Technical Report, 14–5*, 1–9.

Haraway, D. (1988). Situated knowledges: The science question in feminism and the privilege of partial perspective. *Feminist Studies, 14*, 575–599.

Harding, S. (1991). *Whose science? Whose knowledge? Thinking from women's lives.* Cornell University Press.

Henrich, J., Heine, S. J., & Norenzayan, A. (2010). The weirdest people in the world? *Behavioral and Brain Sciences, 33*, 61–83.

ICRP. (1975). *Report of the task group on reference man.* ICRP Publication 23. Pergamon Press.

McNeish, J., Borchgrevink, A., & Logan, O. (Eds.). (2015). *Contested powers: The politics of energy and development in Latin America.* Zed Books.

Muggli, M. E., Forster, J. L., Hurt, R. D., & Repace, J. L. (2001). The smoke you don't see: Uncovering tobacco industry scientific strategies aimed against environmental tobacco smoke policies. *American Journal of Public Health, 91*, 1419–1423.

Sanches de Oliveira, G., & Baggs, E. (2023). *Psychology's WEIRD problems* (Elements in psychology and culture). Cambridge University Press.

Schick, S. F., & Glantz, S. A. (2007). Old ways, new means: Tobacco industry funding of academic and private sector scientists since the master settlement agreement. *Tobacco Control, 16*, 157–164.

SDHS Research Report. (2018). *Status of patenting plants in the global south.* https://sdhsprogram.org/document/statusofpatentingplantsintheglobalsouth/

Seed: The Untold Story. (2016). https://www.seedthemovie.com/

Wickson, F., & Wynne, B. (2012a). The anglerfish deception: The light of proposed reform in the regulation of GM crops hides underlying problems in EU science and governance. *EMBO Reports, 13*, 100–105.

Wickson, F., & Wynne, B. (2012b). Ethics of science for policy in the environmental governance of biotechnology: MON810 maize in Europe. *Ethics, Policy & Environment, 15*, 321–340.

Wilhelm, J. P. (2020). *Ethiopian teff: The fight against biopiracy.* DW.com. https://www.dw.com/en/ethiopian-teff-the-fight-against-biopiracy/a-52085081. Published on 21 January 2020.

Part II
Why Science Cannot Ignore Philosophy: Philosophical Bias in Science

Chapter 5
Conflicting Evidence and the Bias that Science Cannot Avoid

I am formally educated in the health and biological sciences and did not become interested in philosophy of science until my early post-doctoral work. I was working on safety assessment of genetically modified proteins that are introduced in the food chain by genetically modified crops. This is when I became familiar with an ongoing scientific disagreement on whether there was enough evidence to conclude safety of such proteins for human consumption. What was striking to me was not only that diverging opinions were based on common facts, but also that new evidence would be welcomed as irrelevant, or as based on a scientifically dubious methodology, from one of the parts. What was meant as 'scientific' or 'unscientific' approach remained however often implicit. Was the scientific approach considered dubious because it was not based on an experiment? Or because it did not collect a sufficient number of observations? Why were different scientists assigning the quality of 'proper science' to different procedures, given that they were all rationally justifiable to my eyes? One could not derive these answers directly from evidence. For the first time I realised that not everything in science can be settled by facts. (Elena Rocca, 2020)

When Experts Disagree, But Not Over Facts

Science is expected to be the most transparent, objective, and bias-free approach available to study reality. This ideal, as well as the faith in science, has sometimes been challenged. We have already seen in the previous chapter how science have been criticised for implicit gender- and racial biases, resulting from and reinforced by data gaps in research where large parts of the global population are excluded or ignored. This criticism has come from both within and outside philosophy, pioneered by feminist thinkers. That there are socio-economical knowledge gaps in science and technology should worry us for reasons of representation, democracy, and relevance. Another worry raised about science concerns the quality of existing research. In his famous paper 'Why most published research findings are false', physician and scientist John Ioannidis argues that many of the results published in medical science are undermined by systematic errors and are likely to be false positives, thus incorrect. The problem he addresses is the fact that many of the scientific results cannot be replicated or supported by further evidence.

Although the type of argument and the boldness of his statement were criticised, there was overall agreement among scientists with the general concern that Ioannidis had raised. After all, Bacon warned us back in 1620 against different types of bias, or dogmas, that scientists might bring with them to their observations (see Chapter 2). In 1979, the founder of modern evidence-based medicine, David Sackett, identified and listed 35 different types of bias that can undermine the credibility of observational epidemiological studies. For instance, researchers might cherry pick the hypotheses that are most easily confirmed by their statistical design. They might even fail to seek confirmation of their results. Sackett recognised the influence of bias and the importance of being aware of any systematic error 'which may distort the design, execution, analysis, and interpretation of research' (Sackett, 1979, p. 51).

In the last decades the scientific community has made substantial efforts to detect, explicate and minimise different types of bias in order to increase the reliability of results from research. One such effort is the *Catalogue of Bias*, which is a collaborative project that created a regularly updated list of various types of bias affecting health research, with proposed steps to prevent them. Despite these endeavours to remove systematic errors in science and thereby increase objectivity, one did not get rid of disagreement. Expert disagreement seems instead to be an intrinsic, natural aspect of science. In their editorial to a *Synthese* special issue on disagreement in science, Finnur Dellsén and Maria Baghramian make the following observation:

> [S]cientific disagreement is unlike many ordinary cases of disagreement in that there is often little reason to think that the disagreement is due to a simple mistake by one of the parties of the type often appealed to in the epistemology of disagreement literature... Rather, if there is disagreement among two or more scientists—or groups thereof—it is most commonly grounded in a more fundamental difference in the methods, background assumptions, or even the scientific outlook—roughly, in Kuhnian 'paradigms'. This suggests that scientific disagreements present philosophers with special challenges that haven't yet been addressed, even in the abstract, in the epistemology of disagreement literature. (Dellsén & Baghramian, 2021, pp. S6012-3)

We have seen in previous chapters that some philosophers of science see plurality of perspectives as necessary for the development of scientific knowledge. According to Donna Haraway, Sandra Harding, Helen Longino, and many others, we should accept and even embrace the plurality of background assumptions, theories, and research programs. There is a practical difficulty, however, when expert disagreement slows down urgent decision-making processes, and—at worst—harm public trust in science and science-based decision-making. When experts interpret the same scientific data differently and disagree over how to weigh the same pool of evidence, even when a reasonable amount of data is collected, they might arrive at entirely different conclusions. One such case of scientific disagreement is over the herbicide glyphosate. Back in 2016 the International Agency for Research on Cancer (IARC) classified the herbicide glyphosate as a 'probable human carcinogen', while just a few weeks later the European Food and Safety Authority (EFSA) concluded that '… glyphosate is unlikely to pose a carcinogenic hazard to humans' (Portier et al., 2016). Should the regulations around glyphosate have been kept untouched, or should they have been modified?

Science based decisions are not made easier when one needs to consult experts from different disciplines, which is often the case when dealing with complex matters and wicked problems. Typically, then, the overall scientific evaluation involves complex evidence from different disciplines, where different factors and consequences might be given different weight and priority depending on one's area of expertise. An ecologist can have different concerns and risk tolerances than an economist or a developer, simply because they study different subject matters. When relying upon complex, multi-layered, or even contradictory scientific evidence, decision-makers are left with a problem. Which experts and results should one trust, and according to which criteria should such a choice be made?

Here in part II, 'Why science cannot avoid philosophy. Philosophical bias in science', we will show that there are ways to make expert disagreement more transparent. For this, one cannot simply focus on the fact that there are conflicting results. If the conflicting results are based on the same evidence, one needs to identify other sources of disagreement than the evidence itself. We saw that IARC and EFSA made their opposite conclusions about the carcinogenicity of glyphosate based on the same empirical evidence. And in the story at the beginning of this chapter, we heard that scientists sometimes dismiss some facts and evidence as irrelevant because of the methods that were used to produce them.

Think of the following situation. A certain chemical has been proven toxic in lab animals, or a certain mechanism of toxicity has been identified at molecular level. Yet, such toxicity fails to be demonstrated as significant in humans at population level. How, then, should we interpret this conflicting evidence? Which evidence is more trustworthy, and why? That depends, among other things, on which scientific methods and types of evidence that are accepted within a research tradition.

It might already now become clearer why we need to take expert disagreement seriously, and not regard it as a problem to be reduced, dismissed, or ignored. The consequences of making the wrong decision can sometimes be disastrous, and hard to reverse. Decision-makers should therefore be made aware of non-empirical sources of disagreement, or what we can call 'extra-evidential premises'. Some of these were already presented in part I. Scientists might disagree over what counts as scientific knowledge (Chapter 1), which methods are best to enquire reality (Chapter 2), which questions, theories, and concepts are most central to science (Chapter 3), and to what degree scientific findings can be thought of as objective, neutral, or universally applicable (Chapter 4). Not the least, scientists might have different interests and values. As Sandra Harding states in an interview about her book *Objectivity and Diversity*: 'If people had cared about climate change, we would have had sustainable environments a long time ago. Instead, we have extremely sophisticated weaponry and great computers' (*New Books Network* podcast, 2016, quote at 24.40).

We will now introduce a type of bias in science that we think all researchers should learn about, preferably as part of their science training. It is perhaps the most implicit form of biases there is, and yet, scientists cannot avoid them. As will be explained throughout parts II and III, these biases affect all elements of science, including scientific concepts, theories, methods, and norms of best practice. It should therefore not surprise us that these biases are common sources of expert disagreement and represent a major barrier for genuine interdisciplinarity.

Philosophical Bias in Science

We have seen that experts can arrive at different conclusions from the same empirical evidence. This means that their disagreement is not over the facts, but instead over what significance and scientific value such facts should be given. When this happens, it might be that the disagreement stems from what we have called a 'philosophical bias'; or *b*asic *i*mplicit *a*ssumption in *s*cience (BIAS). In a paper with the title 'Philosophical bias is the one bias that science cannot avoid', we introduced three broad types of philosophical bias (Andersen et al., 2019). Before we continue, it might be worth getting a better idea of each of these types.

Ontological Bias

An ontological bias is a basic implicit assumption in science about reality; how the world is, and what truly or fundamentally exists. This type of bias might be about how to understand some key concepts in science, how to classify things, what are their essences, if any, and what is the true nature of things. One well-known example of an ontological BIAS is René Descartes' mind-body dualism, according to which the mental and the physical are two separate forms of existence. In the health sciences this dualist bias manifests itself in the distinction between mental and physical illness. Psychology and medicine are two separate disciplines, with different theories, methods, practices, and institutions. This dualist bias can be challenged, of course. Some might think that the mind is nothing more than the workings of the physical brain, which is a bias of reductionism or materialism. Another alternative to dualism is holism; the idea that the mind and body are not two separate existences, but instead must be understood as an integrated whole. All these three ontological biases—dualism, reductionism, and holism—are represented in the health sciences and in science in general, and motivate conflicting ideas about which theories to accept, how to understand central concepts, and about what counts as best practice.

In the following chapters we will look at several examples of ontological bias in science, including biases about complexity (Chapter 7) and causality (Chapter 8). Here is a case related to both. When James Watson famously described DNA as the code for life, he thought that the structure and functioning of DNA would ultimately allow us to understand life itself. That molecules are the source of life and relations, however, is not an empirically observable fact. Rather, it's an assumption that causes travel *bottom-up*, from the lower micro-level to the upper macro-level of complexity. Such bottom-up causality, however, is not the only way to conceptualise molecules and life. One can also hold the opposite ontological bias, of *top-down* causality.

This idea would entail that complex higher-level relations influence and shape the development of life and molecules at the lower level. Following this ontological assumption, if we wanted to understand genes and their functioning, we would first need to understand the relationships that shaped them. The bottom-up assumption is compatible with a worldview where genes and molecules have stable properties and essences that determine how they interact. In contrast, top-down causality is a more relational worldview, which assigns a fundamental role to interaction and co-production. Bias of an ontological type will affect scientific theory and concepts, but also how a phenomenon is studied.

Epistemological Bias

An epistemological bias is a basic implicit assumption in science about knowledge; how we gain knowledge, what the most reliable sources of knowledge are, to what degree knowledge is influenced by us, or what the highest form of knowledge is that science should aim to achieve. This type of bias will motivate how we develop research methods, how we favour certain methodologies over others, and how we think about the knowledge that is produced by these methods. One might also dismiss some methods as flawed, biased, or useless if they are not fit to pick out the type of information that is seen as most important. Epistemological bias can be closely linked to ontological bias, therefore, since one might have certain worldviews and ideas about the true nature of things that again requires certain methods. For instance, if we think that the mind is best understood as a manifestation of physical processes in the brain, then this will have implications for how to study the mind and which tools are expected to give relevant information. Recall for instance that according to the field of neurobiology, one way to get scientific insight about the origin of emotional response is to monitor the activity of different brain parts through a fMRI scanner (see Chapter 1), something that is meaningful for those who hold a reductionist view of emotions.

We have seen in Chapter 1 how empiricists and rationalists disagree over what counts as the most trustworthy type of knowledge. Empiricists, such as John Locke and David Hume, would say that we can only know what we can observe, which is a common epistemological bias in science. From this perspective, one would place much more emphasis on empirical data than on theoretical explanations or laws of nature. Theories that are not based in solid data would then have little value for someone with this type of bias. This discussion goes back to Aristotle and Plato, where Aristotle was more of an empiricist who was interested in understanding the changing and material world, while Plato was a strict rationalist who thought that we should search for the unobservable principles behind the messy reality. Knowledge, if worth anything, should according to Plato be universal, abstract, ideal, and independent of time and place. In science, the rationalist bias means that one would be more interested in universal laws and general principles that can explain the behaviour of things, than in collecting data from natural settings. One might have to physically create artificially closed conditions in order to establish such law-like behaviour, or one could stipulate universal laws and principles within a theoretical model under some ideal conditions. Theoretical physics is more concerned with laws of nature and generalisations, and in line with Plato, mathematics is one of the most important scientific tools. It should be clearer from this why bias of the epistemological type can have major influence on scientific method, thinking, and practice.

Ethics Bias

The third type of philosophical bias is normative, which would be basic implicit assumptions about ethics, norms, and values. Although scientists are used to evaluate and declare their own conflicts of interests, many values and priorities enter research unnoticed. Scientists might not even think about the various values that underly a research agenda, choice of method, or selection of data. Often such values are easier to detect by someone who disagrees with them, and this is how values and interests are often discussed within the scientific community, as linked to personal biases. There are, however, some ethical standards and premises that are shared widely within the research communities, which makes them much harder to spot. These are the basic implicit assumptions that can be traced back to certain philosophical positions within ethics. Many scientists might share a utilitarian bias, for instance, with a focus on minimising negative consequences and maximising positive ones. In contrast, someone with a duty ethics bias would tend to focus on moral principles and universal rights, rather than on consequences. This type of bias could reveal itself in a discussion on how to frame a research problem.

Another example of ethics bias in science are the different value hierarchies that can be found in disciplines or research agendas, where some populations are given more scientific value than others. As we saw in Chapter 4, one might detect a male or Western-centric bias in some areas of research, where these populations have been given priority in framing which research problems to address, but also in data collection, or in the evaluations of consequences and risks. There are also commonly accepted value hierarchies applied to animals and nature, where certain species are ranked higher than others in certain disciplines. An ecologist might for instance give more value to red-listed species, because they are in danger of extinction, than to black-listed species, which are invasive species that are considered a threat to the local eco-system. Animal hierarchies are many and varied, and they motivate expert disagreement over conservation, protection, and regulation involving farmed animals, wild animals, research animals, and pets. The most common value hierarchy in science is anthropocentricism, where one values human life and welfare more than everything else, and where animals and nature are seen as valuable insofar as they are important to us. Figure 5.1 shows a common value hierarchy, where humans are above animals, with living and non-living nature at the bottom. Alternative value hierarchies to anthropocentrism are: zoo-centrism, where humans and animals have equal value;

bio-centrism, where all living organisms have equal value; and eco-centrism, which also gives equal value to non-living nature.

One extreme case of anthropocentric decision-making was when 17 million minks were slaughtered in Denmark in 2020 to prevent further spreading of the corona virus from the animals to humans. At that point, 12 people had been infected with a mutant strain of the COVID-19 virus from minks, which is what motivated the controversial public health decision. The minks were killed regardless of whether they were contaminated or not, something that reveals an anthropocentric utilitarian bias, according to which the end justifies the means.

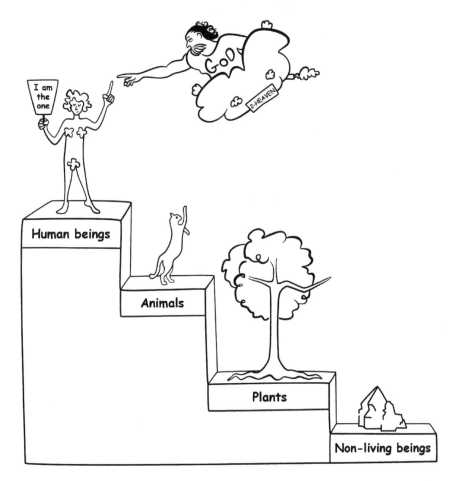

Fig. 5.1 A common anthropocentric value hierarchy, placing humans above all else (Illustration by Sheedvash Shahnia©)

When Is a Philosophical Bias a Regular Bias?

We use the term 'bias' as an acronym for 'basic implicit assumptions in science'. But could a philosophical bias also work as a regular bias? Absolutely. Just like the other types of systematic errors listed by Sackett and his collaborators, a philosophical bias can skew the development of hypotheses, design of experiments, evaluation of evidence, and interpretation of results. Crucially, however, a philosophical bias function as a regular bias only when these basic assumptions remain *implicit*, and the scientists remain *unaware* of their own basic assumptions and the ways these influence their research. This is not always the case, as should be clear from the following example.

Determinism is a philosophical assumption that many scientists share, that there is only one possible outcome for any given set of initial conditions. This assumption makes predictions possible, unlike indeterministic cases, where the same set of initial conditions could result in at least two different outcomes. A coin toss can be interpreted as a deterministic or indeterministic situation, depending on one's philosophical bias. In the deterministic interpretation, the outcome of the coin toss is thought to be in principle predictable from the initial conditions of the toss, so either 1 or 0 probability of each outcome. Understood as an indeterministic matter, however, the outcome would be genuinely probabilistic, so close to 0.5 for each outcome.

What, then, if a scientist purposefully and explicitly assumes determinism in a model to predict population growth? In the model, the population density is at every moment determined only by the initial density. Is determinism here a bias? Not really. Because it is an explicit methodological choice, and not an implicit premise, it doesn't function as a bias. When an assumption is adopted as a helpful premise to investigate reality, it is possible to consider critically how it affects the results. For instance, the researcher might want to check how the model prediction differs from reality, and thus collect evidence about the factors that influence population growth other than initial density of population. A basic, but *explicit*, assumption in science is thus not a bias.

A philosophical bias functions as a lens through which scientists see new information, and they influence how relevant evidence is produced and evaluated in research. As such, philosophical basic implicit assumptions work in the same way as other biases in science. Still, there is an important difference between what we have called a philosophical bias and other biases in science. While it is generally accepted that bias is a flaw that one should try to minimise or eliminate, this isn't possible for philosophical biases. They are an indispensable premise for any scientific activity. We cannot make no basic assumptions at all. For instance, if one disagrees with the philosophical idea of mind-body dualism, it helps to know whether this is because one's own philosophical bias is in line with holism or with reductionism. This is an ongoing philosophical discussion within the health sciences, and regardless of which side one settles on, there is a corresponding philosophical position. This is why we say that a philosophical bias is the one bias that science cannot avoid.

Why Philosophical Biases Should be Made Explicit

Philosophical biases usually get passed on from one generation to the next via educa-tion and research practices. These basic assumptions enter almost every aspect of research: theories, concepts, methods, research agendas, and everything else that Thomas Kuhn described as a scientific paradigm. Normal science is then a result of implicit consensus about these matters, but also about basic philosophical assump-tions. This is why Kuhn saw it as a sign of crisis in paradigm when scientists start engaging in philosophical discussion.

> It is, I think, particularly in periods of acknowledged crisis that scientists have turned to philosophical analysis as a device for unlocking the riddles of their field. Scientists have not generally needed or wanted to be philosophers. Indeed, normal science usually holds creative philosophy at arm's length, and probably for good reasons. To the extent that normal research work can be conducted by using the paradigm as a model, rules and assumptions need not be made explicit. (Kuhn, 1962, p. 88)

The idea that normal science involves a sharing of philosophical and other basic assumptions seems plausible. If one disagrees over what counts as science, which research methods are best, which theories to accept, and what research problem one is trying to solve, it's not easy to work together and pull in the same direction. If one also uses the same concepts with different meanings, one will talk past one another, and could easily arrive at divergent conclusions. One challenge of interdis-ciplinary collaborations is that one might not share most of these basic assumptions. Often researchers from different disciplinary backgrounds will have adopted different philosophical biases through their training. For instance, ecologists might not concep-tualise genes in the same way as molecular biologists. In molecular biology, genes are typically depicted as prone to isolation and manipulation. In ecology, genes tend to be conceptualised as dynamic and interactive. This type of tension might not be a noticeable problem in most discipline-specific research, but philosophical bias represents a barrier for interdisciplinary collaboration and communication.

What, then, can be done about philosophical biases? The first step is awareness that these basic implicit assumptions exist in scientific thinking and practice. Next, one should identify one's own philosophical biases, and those predominant in one's field, education, and research framework. The next step is to discuss these philo-sophical biases explicitly, and to include them in scientific argumentation, teaching, and communication. As part of the process, one also needs to learn about some alter-native basic assumptions, so one can discuss strengths and weaknesses of each of these and make comparisons. A good scientist, we think, is not only someone who knows how to produce relevant facts, but someone who can also motivate why such facts are relevant in a self-critical way.

There are three reasons why we think that philosophical bias should be explicitly stated and discussed. First, identifying an extra-evidential source of disagreement is useful to gain a better understanding of the competing perspectives. This is espe-cially important since we are increasingly faced with complex global challenges that require interdisciplinary approaches. It's no longer sufficient in the post-normal era that scientists are well trained in their own areas of expertise. They also need to be able to communicate with scientists from other disciplines, understand their perspectives, and detect some foundational sources of interdisciplinary disagreement. Explicating philosophical bias can then break down some barriers for interdisciplinary collaboration and communication. Secondly, we think that not only scientists, but also science-based practitioners, would benefit from a philosophical analysis of their own activity. Practitioners could benefit from learning about their own philosophical biases, and possible tensions between these and other dominant biases in their profession. This could facilitate a reflection about the norms for best scientific practice, what types of evidence is considered more reliable and why, and how to make decisions in light of conflicting or partial evidence.

A third reason why we think that philosophical bias should be explicated concerns science-informed decision-making. There are many advantages of allowing a plurality of perspectives in science, as we saw in Chapter 4, but such plurality can also be a problem for the decision-making process. Often, one will have to favour one opinion among others, and for this some guide or criteria are needed. In such cases, it can help to analyse the extra-evidential assumptions involved and make them transparent and open for critical examination. Many philosophers of science, including Helen Longino, Donna Haraway, and Sandra Harding, have been arguing for an explicit discussion of values to help promote democracy in science. Heather Douglas is another prominent voice in this context. She argues that, since evidential underdetermination is a natural part of science, we should expect that expert groups evaluate evidence differently. According to Douglas, acknowledging the extra-evidential aspect of scientific evaluations opens to stakeholder participation in *all* processes of science and therefore favours democracy in science. This, we think, is an aim worth promoting, and a good reason for allowing the wider community of stakeholders of science to take part in these discussions.

Who Cares About Philosophical Bias? And Why Should They?

Of course, not everyone will see the point of engaging in this type of philosophical reflection. If one anyway cannot get rid of philosophical biases, why bother? At a first glance it might seem like a futile exercise in abstract thinking, especially for those who have no training in the humanities. Despite this, we have experienced that many respond positively when introduced to these deeper issues; not only philosophers, but also scientists, practitioners, students, and other stakeholders of science. So, who do we think could benefit from learning about philosophical biases, and why should they be motivated to do so? We will now share some of our own experience with teaching philosophical bias to non-philosophers over the last years, within and outside of academia.

Why Researchers Care About Philosophical Bias

Much research today is expected to be interdisciplinary but, as we have explained, different disciplines and research traditions come with their own set of philosophical biases. This can be a major barrier for genuine interdisciplinarity and communication of research. Research traditions typically determine the norms of best practice, the understanding of key concepts, the preferred tools and methods, and even the values and priorities. As a result of diverging standards, there is often a lack of understanding and even respect across research traditions, where alternative practices are dismissed as poor-quality research or unscientific. When different disciplinary contributions are required for research, it's therefore easier to simply divide the various project tasks into disciplinary chunks. If so, interdisciplinary communication and collaboration is not really needed to carry out the different parts of the project. This, however, is not we would call genuine interdisciplinarity.

In our experience, researchers find it helpful to learn how philosophical bias can influence research practice, and why someone could have other basic implicit assumptions from them which motivate different research practices. First of all, this gives researchers a better understanding of their own disciplinary and personal biases, and thereby of what ideas motivate their scientific standards, ideals, and practices. More importantly, perhaps, is that they realise that different standards, ideals, and practices can be equally rational and scientifically sound. They might also see the added value of combining diverse research perspectives and approaches when trying to understand and solve complex problems. For researchers to see the value in learning about and discussing philosophical bias, it helps to apply them to relevant cases of scientific controversies. In part III, we analyse some controversies for philosophical bias, to illustrate how it can be done.

Why Practitioners Care About Philosophical Bias

One thing is that researchers can benefit from learning about philosophical bias, which we might expect. More surprising to us was how easy it was to get practitioners interested in these foundational issues. In our CauseHealth project, which initially consisted of an interdisciplinary network of philosophers, medical researchers, and healthcare practitioners, we looked at philosophical biases underlying the norms of best evidence and practice. Clinicians and other healthcare practitioners got increasingly involved, and the network continued to expand. To be given the opportunity to reflect critically upon the premises for their own profession, and to gain a better understanding of where these came from, was something that had been missing. In particular, health practitioners were interested in the different scientific ideals underlying so-called person centered and evidence-based practices, but also in issues related to causal complexity, reductionism, and chronic illness. Because many of them had experienced a tension between treating their patients as unique individuals and the statistical methods used to generate evidence from certain populations, they were eager to learn more about ontological and epistemological assumptions that were not much discussed in their profession. Contributing to increased awareness and understanding of these foundational issues, as well as learning a language for discussing them, was welcomed as informative and empowering.

Many of the CauseHealth clinicians and researchers have continued to communicate and spread these ideas and reflections within their own profession, and have even developed new resources for health professionals to learn more about philosophical biases. Some say they wished philosophy of science had been part of their formal training. Others say that methodological and ethical discussions with their peers have become more constructive, because they understand better the rationale behind the choice of different methods and practices. Most importantly, we learned that also patients have an interest in knowing about philosophical bias, since they are the ones affected by the standards for healthcare practice.

Why Students Care About Philosophical Bias

Inspired by the engagement from researchers and practitioners in discussing philosophical bias, we developed a pilot teaching course open for all students at the Norwegian University of Life Sciences (NMBU). Since NMBU is an interdisciplinary university with a specific focus on sustainability research, we knew that the students would be exposed to a plurality of perspectives and standards in their courses and programs, with corresponding philosophical biases. The pilot course was called 'Interdisciplinarity and expert disagreement on sustainability', advertised to teach students about philosophical bias as a source of expert disagreement and scientific controversy. 70 students signed up for the course, and in the end, the class wrote an open letter to the university leadership: 'A call for promoting critical thinking for interdisciplinarity in NMBU'. Here is an extract of the letter:

> The course gathered a wide range of students, from first year bachelor pedagogics to PhD scholars in biology, in engaged and fruitful dialogues and discussions on sustainability issues. We learned to recognize our own biases and communicate them explicitly to students that are coming from other disciplines. This transforms disagreements into a constructive dialog and creates potential for scientific collaboration. For all of us, it has demonstrated the massive need for awareness and transparency around the different philosophical basic implicit assumptions that underlie scientific thinking. Only when investigating the fundamentals of our scientific traditions, we can see past generic jargon and methods to understand how different disciplines reach fundamentally different scientific conclusions and results from the same set of evidence. We, the young and aspiring generations, see the future as fundamentally uncertain, and view the complexity of our societal challenges with great concern. We are happy that the university has chosen interdisciplinary research and interdisciplinary studies as solutions to our imminent sustainability challenges. Now we believe it is the university's responsibility to develop a culture for transparency and critical discussion on conceptual premises and meta-empirical issues that facilitates interdisciplinarity. We ask that a course on expert disagreement and philosophical biases has to become a fundamental part of all bachelor and master programs to create awareness about discipline specific biases.

In addition, there should be more specific (elective) courses to experience, train and practise key interdisciplinary communication skills based in critical thinking. ...This is the education we want and need to be able to face our imminent challenges. (Tuntreet, 2020)

Today the course is offered twice a year at the university, mandatory for several master programs in environmental sciences and natural resource management. Students reported back to us that the conflicting premises and standards used by teachers from different disciplines were previously hidden to them, leaving them with the difficult task of interpreting and trying to make sense of the diverging messages, but without having the tools to do so. Some students have later told us that the acquired knowledge about philosophical bias has been valuable for their subsequent studies and jobs, where interdisciplinary competence was needed.

Is Any Philosophical Bias as Good as Another?

We have argued that philosophical biases should be identified, explicated, and critically discussed by researchers, practitioners, and students. We have also said that one cannot get rid of philosophical biases altogether. The only way to lose one bias is to replace it with another basic, but preferably *explicit*, assumption. Is there, then, some ways to choose between competing philosophical biases, or are they all equally good? Given that centuries of philosophical thinking did not solve these issues yet, how can researchers, practitioners, and students be expected to do a better job deciding which basic assumptions one should adopt? There is disagreement about this point.

In 'Making sense of non-factual disagreement in science', Naftali Weinberger and Seamus Bradley warn against a reinforcement of methodological silos and aggravated disagreement as a result of making philosophical bias explicit. At worst, one might simply give scientists a philosophical reason to leave their position unchanged. We agree that although transparency about philosophical bias might be necessary for understanding expert disagreement in many cases, it's not alone sufficient to resolve disagreements. We might also need some tools to compare and evaluate opposing philosophical assumptions, especially for the purpose of decision-making. Ultimately, we need to be able to say that, among the many rationally defendable policies available, one of them is preferable given the current knowledge. How could this be done in practice?

One proposal is that, when faced with several equally valid evaluations of the same evidence, one should prefer the evaluation whose underlying bias fits better with the wider field (see Andersen & Rocca, 2020 for details). According to the proposed strategy, the first step is to detect and analyse the philosophical assumptions underlying the different positions in a scientific controversy. This can be a demanding task that might require contributions from philosophers, as we illustrate in part III. Once the philosophical biases are explicated, the next step is to compare assumptions in the specific case with the ontological assumptions in the relevant field of research. What do the cutting-edge knowledge and practices suggest? Is the field progressing toward one bias and away from another? This is a complex question that requires consideration of the dominant arguments in the scientific community and the trends in practice.

Does this mean that one should automatically adopt the philosophical bias that is the majority view in science? Clearly not. Historically, many innovating ideas in science have started out as the minority view. The challenge with competing theories, methods, and practices can therefore not be resolved democratically or philosophically. Philosophical analyses of such controversies can nevertheless be helpful. There are clear advantages of applying philosophical analyses for the purpose of science-based decision-making. On the one hand, it encourages development of plural perspectives within basic research and the philosophical transparency of such perspectives. On the other hand, it also highlights the philosophical state of the art of a particular field at a particular point in time. As new perspectives emerge and gain traction, policy and decision-making will follow. Such awareness of how philosophical biases change within science can contribute to a more dynamic, self-critical, and philosophically reflected development in both basic research and policy.

Reviving the Discourse on Basic Assumptions in Science

The idea of a philosophical bias is inspired by a long-lasting tradition dating back to the scientific revolution. Already then, it became clear that every scientific theory comes with a set of implicit philosophical assumptions and that it's possible to construct different models and interpretations of a single data set. Philosophers use the term 'underdetermination' to describe cases where the same data can motivate and support different theories. In this chapter, we have tried to show how the issue of evidential underdetermination is relevant also in modern scientific controversies, especially for socially relevant science and for science-based decision-making. Our aim is to emphasise that attention to basic implicit philosophical assumptions in science has relevance beyond philosophy itself. At best, awareness of philosophical biases can be a tool for genuine and constructive transdisciplinary discourse and decision-making. In the following chapters, we will introduce some ontological, epistemological, and ethical biases that enter all types of research. Does science uncover or construct truths (Chapter 6)? How to understand and analyse complexity (Chapter 7)? What is the nature of causality, and what are the best methods for establishing it (Chapter 8)? How is risk and risk assessment related to value and to our understanding of probability (Chapter 9)? By discussing these selected concepts, we hope to introduce the reader to the foundational role of basic assumptions and show the relevance of philosophy for science and science-based decision-making.

Chapter Summary

We started by pointing out that experts still disagree over the same set of evidence, despite of an increased focus on objectivity and the effort to remove bias for the scientific process. We have given our own account of expert disagreement by saying that it can derive from divergent philosophical basic implicit assumptions in science (philosophical bias) which can be of an ontological, epistemological, or ethical nature. Differently from other types of bias, that can be considered as systematic errors, philosophical biases are essential for the scientific process and thus cannot be removed. However, we propose that philosophical bias should be identified and discussed explicitly in order to transform disagreement into constructive dialogue. We have listed some reasons why researchers, practitioners, and students of the natural and social sciences care about conceptual reflections of philosophical bias. These are arguments that we have encountered in the last years during our research and teaching activities, and we have found them compelling. They encouraged us to advance the knowledge on the topic.

Further Introductory Reading

For a further introductory discussion on philosophical bias, including more examples, see our 2019 article 'Philosophical bias is the only bias that science cannot avoid', with Fredrik Andersen. For an introduction to an example of philosophical bias in the field of drug safety, see 'Philosophy of science meets patient safety' by Elena Rocca (2020). If you are looking for a broader introduction to the field of disagreement in science, see the special issue on 'Disagreement in science' published by the journal *Synthese* and edited by Finnur Dellsén and Maria Baghramian (2021). If you are interested in the question of objectivity bias in science, you might find it interesting to look at the website *Catalogue of Bias* curated by the Oxford Centre for Evidence-Based Medicine, as well as the much-discussed article by John Ioannidis (2005), 'Why most published research findings are false'.

Further Advanced Reading

Heather Douglas (2012) discusses the role of philosophical bias about ethics when different scientists and decision-makers evaluate the same set of complex evidence in 'Weighing complex evidence in a democratic society'. For the detailed presentation of a method for identifying implicit basic assumptions in a scientific controversy, see 'How biological background assumptions influence scientific risk evaluation of stacked genetically modified plants' by Elena Rocca and Fredrik Andersen, (2017). The same authors also offer an advanced description of the strategy for selection of one philosophical bias for the purposes of decision-making in 'Underdetermination and evidence-based policy' (Andersen & Rocca, 2020).

Free Online Resources

In the digital conference *Interdisciplinarity, Sustainability and Expert Disagreement* (2020), we invited philosophers to present expert disagreement in different fields related to sustainability and to analyse them in terms of the underlying philosophical biases. The conference website is still available, and all the video presentations are open. All talks have written discussion sessions that can be found in the chat following the video.

Study Questions

1. What is a philosophical bias, and how do they contrast with other biases? How do you think a logical positivist would respond to the idea of philosophical basic assumptions in science?
2. Are you familiar with any philosophical biases from your own education or discipline? Do you share any of these biases?
3. Give some examples of ontological, epistemological, and ethical bias in science.
4. How can philosophical bias affect science?
5. How is it possible for scientists to agree over empirical facts, but still disagree over which conclusions to draw from those facts?
6. Why should scientists be aware of philosophical bias, and discuss them explicitly? What do you think about the idea that scientists should engage in philosophical discussion of this type?
7. Which other groups in society do you think might benefit from knowing about philosophical assumptions and their influence on science?
8. How do you think one should choose between two opposing philosophical positions or bias, if there are rational grounds to adopt either?

Sample Essay Questions

1. Present a scientific controversy that is not primarily over facts, but basic implicit assumptions of a philosophical nature (ontology, epistemology, or ethics). Explain how different philosophical biases motivate different positions in the controversy. Make sure you also explain what a philosophical bias is and how they differ from other biases.

2. Different disciplines might assume different value hierarchies, where humans, animals, nature, or even certain species are ranked above other. Present and discuss some examples of value hierarchies in your own education or discipline: anthropocentrism, zoo-centrism, bio-centrism, or eco-centrism. Make sure you include your own perspectives on value hierarchies.
3. Present the idea of a philosophical bias and explain their role in science. Discuss the claim that scientists and others ought to engage in discussions about basic implicit assumptions in science, including your own thoughts on the matter.

References

Andersen, F., & Rocca, E. (2020). Underdetermination and evidence-based policy. *Studies in History and Philosophy of Science Part C: Studies in History and Philosophy of Biological and Biomedical Sciences, 84*, 101335.

Andersen, F., Anjum, R. L., & Rocca, E. (2019). Philosophical bias is the one bias that science cannot avoid. *eLife, 8*, e44929.

Dellsén, F., & Baghramian, M. (Eds.). (2021, November). Special Issue on disagreement in science. *Synthese, 198*(25), 6011–6021.

Douglas, H. (2012). Weighing complex evidence in a democratic society. *Kennedy Institute of Ethics Journal, 22*, 139–162.

Interdisciplinarity, Sustainability and Expert Disagreement. (2020). Online conference. https://int erdisciplinarityandexpertdisagreement.wordpress.com

Ioannidis, J. P. A. (2005). Why most published research findings are false. *PLoS Med, 2*(8), e124. https://doi.org/10.1371/journal.pmed.0020124

Kuhn, T. (1962). *The structure of scientific revolutions.* University of Chicago Press.

Portier, C. J., Armstrong, B. K., Baguley, B. C., et al. (2016). Differences in the carcinogenic evaluation of glyphosate between the International Agency for Research on Cancer (IARC) and the European Food Safety Authority (EFSA). *The Journal of Epidemiology & Community Health, 70*, 741–745.

Rocca, E. (Podcast 2020). Scientific disagreement and philosophy. *PedPod—NMBU Pedagogy Podcast*, Episode 4. https://www.nmbu.no/ansatt/laringssenteret/kurs-og-kompetanse/pedpod

Rocca, E., & Andersen, F. (2017). How biological background assumptions influence scientific risk evaluation of stacked genetically modified plants: An analysis of research hypotheses and argumentations. *Life Sciences, Society and Policy, 13*, 11.

Sackett, D. (1979). Bias in analytic research. *Journal of Chronic Diseases, 32*, 51–63.

Tuntreet. (2020). Open letter: A call for promoting critical thinking for interdisciplinarity in NMBU. *Tuntreet* (NMBU student newspaper), *6*(75). https://issuu.com/tuntreet. Published on 10 September 2020.

Chapter 6
Does Science Uncover or Construct Truths? Bias about Observation

Is the Ideal of the Unbiased Researcher Itself a Philosophical Bias?

We have introduced the idea of philosophical bias in science. While other biases can and should be avoided, we said, a philosophical bias can only be replaced with another bias or be explicitly stated as a premise for research. There is, however, a view that even to talk about bias in science, and to promote an ideal of bias-free research, suggests that one is already committed to a specific philosophical bias. If so, it seems that talk about 'bias' is itself a sign of a philosophical bias. The concept of bias suggests that science could, in principle, be entirely neutral and free from systematic errors, but also from methodological and philosophical assumptions, values, and other presuppositions. The aim of research could then be to uncover truths about reality, as they are, independently of us. If the results are influenced or contaminated by the researcher, that would amount to bad science.

The point that the idea of bias is itself a philosophical bias was first proposed to us by our friend and colleague Svein Anders Noer Lie, who argues that many philosophers and scientists share what he calls a *non-relational* ontology. According to this view, the phenomena that scientists might want to describe are thought to exist independently and unrelated from everything else, and as possessing their properties intrinsically, independently, and essentially. Only under this assumption can one get the idea of the unbiased researcher as a scientific ideal, where the researcher should avoid adding anything of themselves to the phenomenon. The main role of the researcher is then to uncover the properties as they truly are, intrinsically and unrelated to us and everything else. In contrast, Lie (2017) argues for a relational ontology, according to which any such independence is impossible. This would mean that the idea of independence is a bias. In other words, he suggests that *the ideal of the unbiased researcher is motivated by a philosophical bias*. A consequence of this, he argues, is that scientists who start from a different philosophical point of view in their research might be thought of as unscientific or biased. This is why it is vital

R. L. Anjum and E. Rocca, *Philosophy of Science*, Palgrave Philosophy Today, https://doi.org/10.1007/978-3-031-56049-1_6

that the scientific community is aware of and engage in discussions about these basic philosophical assumptions.

In Chapter 1 we presented different philosophical views on what type of knowledge deserves that name science. Recall that rationalism is the idea that true knowledge comes from reason and abstraction, while empiricists see sense experience as the most reliable source of knowledge. A third view is perspectivism, according to which all knowledge is situated and somehow relative to a perspective. We will now see how different epistemological perspectives motivate different ideals within the natural and social sciences. Specifically, we will look at some ideals for objective scientific knowledge, the nature of observation, and the role of the scientist. One view is that the scientist should be a neutral observer, aiming to uncover truths about the world that are untainted by their own biases. A contrasting view is that all scientific truths are mental or social constructs, and that the role of the researcher is to be an active participant and co-creator of knowledge. We present different philosophical stances between these two opposites, and explain how they translate into different views on the reliance on data or theory in research, and which, if any, comes first in observation: data or theory. We start by presenting the two opposing views where the contrast is most visible, also to someone unfamiliar with the debate.

Science Should Aim to Uncover Truths

A common scientific ideal is that the aim of research is to uncover truths about the world that are factual, objective, and unbiased. When correct, scientific theories describe and explain facts about real-world phenomena, as they are independently of us. At first glance, the alternative seems implausible or at least unattractive. If scientific truths were *not* objective facts about the world, but just a reflection of our own subjective experiences and interpretations, then why should we trust them? Indeed, what would we need scientific theories and results for, and to what might they apply? Could one even meaningfully say that a 'truth' is subjective, or would that take away the whole meaning of truth? Surely, we are not really interested in learning about the personal views, biases, and interests of individual researchers. What we want from science and research is to learn something objectively true about the world. If not, why would anyone even bother with science?

In Chapter 1 we saw how the logical empiricists (positivists) argued that science ought to be a purely empirical matter. That way one avoids mixing in one's own expectations or biases. According to the positivists, the researcher should stick to what can be observed, and they were critical of any claims that cannot be confirmed by observations. Claims about values, interests, religion, or philosophy have no place in science, only factual claims. There are two important implications of this for what exactly research ought to be and how it should be conducted. One side of this positivist bias is what becomes the primary object of research, namely observable facts. This would fit David Hume's strictest form of empiricist philosophy, according to which anything that goes beyond the data should be considered metaphysical speculation.

For the positivist, a claim that isn't empirically justifiable is not only speculative, but *unscientific*.

Another side of the positivist bias is how the role of the researcher is primarily to be a meticulous and unbiased observer. Recall that Hume thought of the human mind as initially an empty slate, or bucket, that gradually gets filled up with more and more sense impressions. The fuller the bucket is, the more knowledge one has. Education of experts can then be seen as the process of filling up one's bucket with relevant facts, to later be recollected, compared, and reported by the expert. This is also why one might think that a powerful computer can have knowledge or intelligence. Computers can store enormous amounts of information and recollect it in less than seconds. If such recollection and reporting were to be tainted by bias, we might think of the information we receive as contaminated, flawed, and untrustworthy. Ideally, one might think, the information reported should be uninterpreted, or 'raw', which is an important ideal in data science and machine learning.

These different scientific and philosophical ideals are related. Taken in combination, they reveal a fundamental assumption about how we interact with the world we live in, as observers and recorders of empirical facts. On Hume's empiricist view, the observer plays a more passive role, where the external world is causally affecting us, but not vice versa. Any knowledge we have must first have entered our minds as sense data, and we in return should add nothing of ourselves into it. It is, so to speak, a one-way causal path from the external reality to our minds via our senses: sight, hearing, smell, touch, and taste (see Fig. 6.1).

If we accept the ideal of strict empiricism, then the most valuable type of scientific knowledge we can have would be data from observation, and as much of it as possible. The empiricist's dream is a complete data set, representing all past, present, and future facts. Ideally, of course, these data should be neutral and objective, untainted by the observer. To ensure that science is done correctly, and without bias, it must be done by following a certain procedure. Objectivity is then guaranteed by having a large data set and by using the correct methods to analyse and compare them. Francis Bacon's scientific method offers the perfect procedure for this ideal (see Chapter 2). Science should according to his inductive method start with the collection and systematisation of many and varied data that are completely unbiased. Starting from observation,

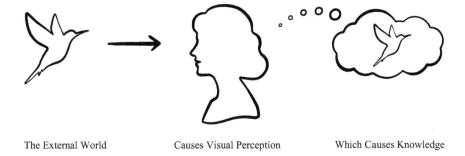

| The External World | Causes Visual Perception | Which Causes Knowledge |

Fig. 6.1 The ideal of the neutral observer. Illustration by Sheedvash Shahnia©

Bacon urged that we in our scientific hypotheses and theories should be cautious about concluding beyond the empirical evidence.

Some philosophers might call this positivist version of empiricism extreme, or even naïve, since it assumes that our observation data can be raw or uninterpreted. In many research traditions, however, this is a common ideal. According to this ideal, scientific truths should be primarily about those data. Without data, one might think, science becomes pure speculation. Theories, in contrast to data, cannot be trusted. This is because theories usually say more than the data and might include explanations of why those data are the way they are. Researchers with a positivist bias tend to avoid suggesting or even accepting scientific theories—unless they simply synthesise the data. From this perspective, therefore, objective scientific truths are guaranteed by sticking to empirical data. It is not the job of science to develop theories to explain those data.

We will return to the relationship between data and theory later in this chapter. Before this, we will present a view about scientific truths that seems to be the radical opposite to what we have described here. This is the position of relativism, or constructivism.

Scientific Truths Are Not Uncovered, But Constructed

There is another view that is equally common in science, but perhaps most prevalent in the social sciences. This view denies the positivist assumption that science should, or even could, uncover factual truths from unbiased observations. According to this perspective, scientific truths and theories are human or social constructs. Our observations will be influenced by us and our expectations, assumptions, and perspectives. As such, scientific observations are subjective, or at best intersubjective and a matter of social convention. This is the philosophical position of constructivism. For the constructivist, there is no such thing as an objective, neutral, or unbiased truth. Science doesn't simply uncover matters of facts; it creates new facts.

In contrast to the positivist ideal, constructivism states that we always influence the world through our prior knowledge, values, and expectations. Instead of the one-way causal influence on us from the external world via our senses, the direction is here reversed (see Fig. 6.2). *We* causally influence the world through our concepts, classifications, and expectations. *We* divide the world into categories, *we* decide what to look for, *we* define and set the standards.

A constructivist would argue that what we take to be scientific 'truths' are created by our perspectives and social cultures. In Chapter 3, we presented some social theories of science, according to which the scientific concepts, theories, methods, and agendas are shaped by the scientific community, or paradigm. Consensus among scientists happens because of a shared set of assumptions and perspectives, and not because one has reached the final objective truth about the world. Constructivism has its philosophical roots in relativism, so it's worth looking a bit closer into this position. Relativism has its origin in ancient Eastern and Western philosophical traditions and

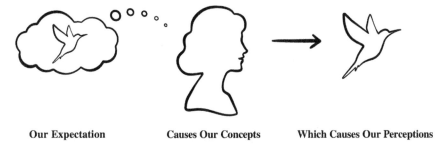

Our Expectation   **Causes Our Concepts**   **Which Causes Our Perceptions**

Fig. 6.2 Observation as influenced by our assumptions. Illustration by Sheedvash Shahnia©

is still a much-debated view. There are two types of relativism. One is epistemic relativism, stating that all knowledge is relative to some perspective or standards. The second is ontological relativism, which says that reality itself is relative to our perspectives or standards.

The earliest known relativist tradition is Jainism, which developed in the period 7–500 BCE in East India. According to the Jain philosophy, reality will always be perceived differently from different points of view. We cannot help but see the world from our own partial perspective, and none of these have epistemic priority over other perspectives. The idea that there can be no single point of view or one side that represents the complete truth is also called 'standpoint theory', and in part I we presented some modern-day versions of this idea.

The early Greek relativists were the sophists, who had noticed that there was plenty of disagreement among experts. All the natural philosophers proposed different theories about the most basic principles of reality. Thales of Miletus proposed that everything in nature has its origin in water, while Anaximenes said it is air, in various densities, that is the basis of all the elements. Around the same period, Heraclitus stated that everything is change. Democritus proposed the view of atomism; that everything in nature is composed of infinitely many tiny, indivisible parts in different shapes. Which theory is the correct one? Relativism was the sophists' rational, but also radical, response to this expert disagreement. With so many different theories that all seem to have some merit, how can one choose between them? Perhaps there are many equally correct descriptions of reality, just like there are different cultural, moral, and religious practices?

How radical should we interpret relativism to be? A famous relativist statement was made by Protagoras: *The human is the measure of all things* ('homo mensura'). This can be interpreted as a strong relativist claim, of subjectivism. According to this version of relativism, truth or knowledge are relative to the individual, and there are simply no criteria for choosing between opposite viewpoints. If Adam says the apple is green and Eve says it is red, then all we can conclude is that the apple is green to Adam and red to Eve. A more moderate version of relativism is conventionalism, stating that truth or knowledge are matters of conventions, for instance of a society or a culture. Still, there might be opposing conventions even within a society, and scientific paradigms could on a relativist interpretation represent such conventions. If so, there are no criteria for choosing between paradigms, as Thomas Kuhn suggested when he said that one cannot compare or translate between paradigms. What is true will then depend on the starting point: which premises and perspectives we bring to the investigation. If so, there can be no truth or knowledge without a starting point, such as a view from nowhere, or everywhere. Relativists would therefore deny that scientific truth can be objective in the sense 'independent' from us, or 'neutral'.

In *Relativism in the Philosophy of Science*, Martin Kusch formulates five principles of relativism. (1) Dependence: knowledge is dependent on some standards (principles, norms, rules, and so on). (2) Plurality: there could be many such standards. (3) Conflict: some of these standards are conflicting. (4) Conversion: switching from one set of standards to another has the character of conversion or a 'leap of faith'. The fifth principle is that the standards are *symmetrical*. This principle can be interpreted as the weaker claim that standards of knowledge are locally grounded (so not universal), or as strong relativism, saying that all such standards are equally valid ('anything goes'). One can replace 'knowledge' with 'moral values', Kusch says, and get a description of moral relativism.

The basic idea is still the same: that all conflicting standards are ultimately grounded in local, variable, and contingent commitments. This is the main problem with relativism according to its critics: that there is no way to settle disagreements and that any perspective becomes equally valid. Perhaps, though, a commitment to relativism or constructivism doesn't have to mean that one must also accept that anything goes ('equal validity'). Indeed, Kusch explains that there are different versions of relativism, stronger or weaker, depending on which of the principles one accepts and how they are interpreted. One might, for instance, accept dependence, plurality, and conflict, but still think that some standards are arguably preferable.

We will now look at some ways in which different philosophical biases about observation motivate different scientific ideals about the role of data and theory. One such ideal is that science should rely on large amounts of empirical data and avoid constructing theories.

The Ideal of Data Centric Science and 'The End of Theory'

Data plays an important role in science and research and for good reasons. But how big is the role of data, exactly? Perhaps science could be based on nothing but data if one just had enough of it? In *Data-Centric Biology*, Sabina Leonelli, a prominent voice in the philosophical discourse on big data, describes the emerging scientific trend as follows:

> Over the last three decades, online databases, digital visualization, and automated data analysis have become key tools to cope with the increasing scale and diversity of scientifically relevant information. Within the biological and biomedical sciences, digital access to large datasets (so-called big data) is widely seen to have revolutionized research methods and ways of doing science, thus also challenging how living organisms are researched and conceptualized. Some scientists and commentators have characterized this situation as a novel,

'data-driven' paradigm for research, within which knowledge can be extracted from data without reliance on preconceived hypotheses, thus spelling the 'end of theory'. …We are not witnessing the birth of a data-driven method but rather the rise of a data-centric approach to science, within which efforts to mobilize, integrate, and visualize data are valued as contributions to discovery in their own right and not as a mere by-product of efforts to create and test scientific theories. (Leonelli, 2016, p. 1)

One reason for avoiding theory and stick to a purely data-driven science, is the problem of underdetermination. Since data don't speak for themselves and tell us which theory explains them correctly, there could be many different theories that could fit the data equally well. Perhaps, though, data are powerful enough to fill all the roles that we need from science.

Let's now consider for a moment how science would look if data did all the work. Data could then be understood as *facts* of reality that scientists can uncover and describe. This might be valuable in itself and, indeed, with enough data one might be able to spot some *patterns*. Say we observe a certain consumer behaviour that suggests a preference toward more environmentally friendly products. This could then *indicate hypotheses*. Is the behaviour motivated by people's moral values, or could it be a response to an increased supply, cheaper products, trends, and advertising? Such data-generated hypotheses could then be *tested* against more data. The results of such tests might give support to the scientific hypotheses, which means that data can also play the role of *scientific evidence*. What type of data is required to count as evidence will depend on the scientific discipline and their preferred methods. It might be enough to have statistical data from observation studies or meta-studies, or one might require data from experiments. In many research traditions it's acceptable or even preferable practice to publish results in the form of data, without adding theoretical or causal explanations. We could even say that the data themselves offer sufficient *explanation*, showing us what actually happened. Data might show that there is a significantly higher rate of lung cancer incidents among smokers who were exposed to radon gas than equally exposed non-smokers. The combination of smoking and exposure to radon could then explain the observed difference. Furthermore, if something happens regularly and across various contexts, we might use these data to *predict* future events. Data can also be *re-used* or *re-purposed* in different research agendas to maximise their utility.

All this taken together suggests that data can provide everything we need to *generate scientific theories*. In order to avoid speculation, however, such theories should be restricted to describing the relevant observations, and the conditions under which they are normally observed. If data can do all this, then it seems that the ideal of data centric science seems quite realistic. The more data that can be shared within the research community, the more knowledge could be generated. By making large data bases openly available, scientists can then harness the insights from other studies in their own research. A question, however, is how well-suited data are for sharing across research agendas, and how dependent they are of the context from which they were obtained. If data are raw, uninterpreted, and independent of a specific research context, then this would be no problem. If, however, data are *constructed*, and tainted by theoretical assumptions; well, then the story becomes quite different.

Data Is Relative to Theory and Context

The clear-cut distinction between data and theory has been challenged by philosophers of science. Specifically, the idea that data can be raw and fit for any purpose has met some criticism. A constructivist would deny that data are simply collected, or uncovered, since any observation would be influenced by the observer. We will now look at some of the objections that have been raised against raw, uninterpreted, and theory-independent data.

Leonelli warns scientists against assuming that data are easily transported from one context to another. 'Data' means 'what is given', but despite of this, she says, data are clearly *made*. They are the result of complex interactions between the researcher and the world and rely on research agendas, theories, concepts, methods, standards, and instruments. Many experiments involve highly advanced machinery and technology to even produce data. One extreme example of this is the CERN Large Hadron Collider, which took 10 years and 4.75 billion US dollars to build even before a single experiment could be done. In what sense, then, could data be thought of as raw, neutral, or free from theoretical assumptions and interpretations?

Leonelli sees data as genuinely relational, local, and situated. Any evidential value they have will depend on the context of inquiry. Researchers should therefore be cautious of using large databases for various scientific purposes without critical consideration. That would be to assume that data remain unaffected by the assumptions, purposes, and methods by which they were generated. But data don't travel that easily, she says. This also has consequences for the mining and curation of these big data sets, she notes, where the choice of classifications and keywords for retrieving the data has enormous influence on how the data will be interpreted and used.

> Irrespective of how standardised they are, the instruments used to generate those data are built to satisfy specific research agendas... This means that we need to acknowledge that no data are 'raw' in the sense of being independent from human interpretation. Moreover, data can be processed differently. It is thus important to understand the conceptual choices that shaped the production and classification of data. Researchers using big data need to recognise that the theoretical structures that informed the production and processing of the data will influence their future use. (Leonelli, 2019, p. 2)

According to Leonelli, data and theory cannot be easily separated. Data are generated within a theoretical framework, and big data mining and infrastructures necessarily adds to the theoretical commitments. In *Philosophy of Open Science*, she argues that scientists ought to consider critically how these and many other philosophical issues enter into the open science movement and its ideals for open data, open access, and so on.

'There is more to seeing than meets the eyeball'

Another philosopher of science who has challenged the clear-cut separation between data and theory is Norwood Russell Hanson. In *Patterns of Discovery*, he argues that any meaningful observation is theory laden. To see something as a particular type of object is according to Hanson to see that it's able to do certain things. Recall the duck-rabbit drawing (Fig. 1.2), where from one perspective it's a duck, but from a different perspective it's a rabbit. Someone who has never seen or learned about ducks might only see the rabbit, and vice versa. Moreover, to see it as a rabbit means that one also sees it as a certain type of animal, with certain properties and behaviours. In this sense, we see what we have learned to see. When observing a phenomenon, therefore, some things might be noticed while other things will be missed.

Hanson uses the example of two microbiologists looking at the same image of a cell but disagreeing over what they see. Where one sees a Golgi body, which is an intracellular organelle, the other sees only traces of foreign matter on the cell, left there in the experimental process. One might, Hanson notes, say that the two scientists see the same thing, but that they interpret it differently. Or perhaps observations are a bit more complicated than that. 'Perhaps there is a sense in which two such observers do not see the same things, do not begin from the same data, though their eyesight is normal and they are visually aware of the same object' (Hanson, 1958, p. 4). He moves on to another example, of scientists Tycho Brahe and Johannes Kepler watching the dawn together. Because Tycho has a geo-centric worldview and Kepler has a helio-centric worldview, they will see this as two different events. Tycho sees the Sun moving, while Kepler sees the Earth moving. Of course, they *see* the same thing in one sense: they both have a retinal response to the sunset. Such retinal responses, however, is not what observation amounts to, Hanson argues. Even an unconscious person, or someone hypnotized, drugged, drunk, or distracted, will have a normal retinal response to the sunset, but they won't actually see it. Seeing is a conscious experience, he insists, which means that it is people who see, not their eyes. The eyeball itself is blind, just like a camera lens is blind. *There's more to seeing than meets the eyeball*, he states.

Hanson explains that a trained scientist can see things that a child or a layperson is incapable of seeing, for instance when looking into a microscope. The child will certainly see something, but to see a Golgi body, red blood cells, or E-coli bacteria, for instance, is only possible for a trained eye. Even with perfect vision, he says, the child is effectively blind to what the expert sees. In order to see what the expert sees, one will have to learn some science. The only sense in which the child and the expert *see the same* in the microscope, is again the retinal response. This, however, doesn't amount to any useful form of observation, suitable as data. The ideal of raw data would then be pointless since they would have no scientific meaning or value. Going back to Leonelli's point about how data are classified and managed, there would be no way to retrieve and reuse data if they weren't already categorised. If 'raw' data means entirely uninterpreted, or empty of meaning, they could play no role in big data science.

An observation will according to Hanson always be dependent on theory. To see something as an X necessarily involves more than simple classification. What does he mean by this? Let's take an example. A small child might correctly identify a bird as a blue tit and thus seem to have an impressive knowledge of birds for their age. However, if they then go on to identify also finches and sparrows as blue tits, we would no longer think so. In fact, we might even doubt that they saw the first bird *as* a blue tit, since they don't seem to know anything about the distinctive features and behaviours of blue tits compared to other small birds. Hanson refers to the informational aspect of seeing as 'seeing that': to see an object as X is to see that further observations are possible, that X might behave in ways that we know them to behave. This adds theoretical knowledge into observation. Science, Hanson says, 'is not just a systematic exposure of the senses to the world; it is also a way of thinking about the world, a way of forming conceptions. The paradigm observer is not the man who sees and reports what all normal observers see and report, but the man who sees in familiar objects what no one else has seen before' (Hanson, 1958, p. 30).

What should we take from this? First, it says something important about the active and participatory role of the researcher in producing data. If seeing must be understood as a conscious activity, then the role of the researcher cannot and should not be as a passive or neutral observer, empty of any presuppositions. That would be impossible. Given that observation data are themselves theory-dependent, and thereby influenced by what we have learned to see, a trained eye would produce better data than an untrained eye. Trying to rid oneself of all presuppositions seems pointless, since one would then have to also erase all prior knowledge including one's entire education. Meaningful data can only come from theoretical knowledge, including concepts and classifications. If 'raw' data means stripped of all meaning and theoretical baggage, they couldn't even be recorded, retrieved, or used.

As we saw in Chapter 2, data will always be produced with a certain method, so will also include methodological presuppositions. We might here recall Nancy Cartwright's point that many laws of nature are *produced* or *constructed* under highly idealised or artificially controlled settings. Only under the whole machinery of science, with labs, tools, and experiments, can we create uniform, law-like, and predictable behaviour out of messy natural phenomena. From this perspective, scientific data are never unbiased, theory-independent, neutral, or raw.

Between Constructivism and Positivism

There could be a middle position between constructivism and positivism, where some aspects of observation are relative to the observer but not everything. For instance, we might accept that the only thing that is relative about science is that researchers share certain standards, tools, and assumptions that they rely upon in their scientific enquiry. If those standards where changed, then one might end up with different research questions, methods, and results. This form of relativism might not be a big threat to the objectivity of science. On the other side, we might accept that it's impossible to do science without assuming anything, such as a theory, certain hypotheses, and some methods. Still, we might not think that this makes scientific knowledge relativist or subjective, or even intersubjective. Perhaps there are very good scientific reasons for having those assumptions.

As explained in Chapter 3, Thomas Kuhn talked about scientific paradigms as the framework that normal science happens within, and where all the research is shaped and motivated by the theories, methods, concepts, research questions, institutions, and authorities that are commonly accepted within that paradigm. In this sense, scientific practice, theories, and knowledge become relative to the shared standards and assumptions within a scientific community, since other standards would result in different research and knowledge. One can accept this, while still claiming that normal science is able to uncover some objective truths about the world. Indeed, even though many of our scientific classifications and dichotomies seem to be strongly influenced by certain assumptions, perspectives, and interests, there are nevertheless some properties in the world that fit certain categorisations but not all.

For instance, there seem to be at least some relevant differences in properties that make some scientifically relevant distinctions plausible, between human and non-human animals, between different plants and species, or between different religions. Whether these are the most important distinctions there are, or to what extent they represent ontologically distinct categories, might of course be debated. But to say that these distinctions are nothing but mental constructs might seem too strong. After all, the world seems to give us enough resistance to prevent us from imposing just any description or categorisation on it. If so, then even though two scientific theories or paradigms can both be perfectly reasonable and offer excellent predictions and explanations, we should expect that there are many such paradigms that are simply wrong and inconsistent with new knowledge or new observations.

Observation can thus be both perspectival and a reliable source of knowledge. One could for instance accept that there are certain matters of fact that cannot be described with whatever scientific theory we want, on the one hand, while also accepting that all scientific knowledge that is produced about these matters of facts is relative to some perspective and assumptions. From this standpoint, one can then combine the ontological view that there are some true facts, with the epistemological view that we can get to know about such facts though plural perspectives, as illustrated in Fig. 6.3. Perhaps Leonelli, Hanson, and Cartwright could all fit within that middle position? Even if both data and theories are constructed, they can still say something true and objective. Cartwright, for instance, argues that scientific results and theories are more *reliable* if they can be backed up by different groups of researchers starting from different assumptions, methods, and research agendas. Research on climate change and global warming is an example she uses (in a podcast) of conclusions that are supported by a wide range of disciplines and research projects, all filling in parts of the picture.

Many standpoint theorists accept this middle position more explicitly, including Sandra Harding, Helen Longino, and Donna Haraway, and also the perspectival realists, such as Michela Massimi. All these philosophers would deny the idea of objectivity as a neutral *view from nowhere*. Instead, they argue that the more perspectives are represented in science, the more objective and complete it becomes. Important scientific virtues would then be to promote transparency and awareness of perspectives, basic assumptions, and value judgements and how they influence research.

Fig. 6.3 One reality, but plural perspectives. Illustration by Sheedvash Shahnia©

This middle view also changes the meaning of objectivity. If data are not raw or independent of various standards and assumptions, then scientists should be transparent about what these are. Objectivity thus comes from transparency about standards and assumptions, rather than from the attempt to exclude them. Scientific evaluations and findings express ideas and values of the scientists—they are at least in that respect *situated*. But, if every evaluation is biased, some biases can still be better than others. From this perspective, the main problem with traditional science is that the background assumptions are ignored or pretended not to be there. As we argued in the previous chapter, researchers should instead acknowledge their assumptions, explicate them, and make them open to critical scrutiny within the scientific community and society.

Natural and Human Sciences and Different Research Ideals

At a conference back in 2016 on expertise, evidence, and argumentation for public policymaking, we heard for the first time the term 'baby quant'. They were referring to the emerging trend of trying to make qualitative studies more quantitative in order to meet certain scientific standards or ideals. What does this mean? Let's say that someone is doing a qualitative in-depth study of a specific workplace, interviewing ten employees. Although the sample is small, one might want to assign scores to the themes identified in some interviews to facilitate comparisons and presentations of results. One might also want to describe numerically and statistically the characteristics of the participants who were involved in the qualitative study. When judged according to the standards of quantitative studies, such baby quant studies will seem of poor scientific quality, for instance because the data sample is tiny and too small to have statistical significance, there is low external validity of the results, and it's not possible to make conclusions about generalisability. On the other hand, if quantitative studies were to be evaluated according to the standards of qualitative research, wouldn't they also seem to be of poor scientific quality? After all, quantitative studies only consider a few variables, and they offer little understanding of contextual influences on the matter. How suitable are statistical studies if we want to understand complex phenomena? If one wants to understand factors that contribute to or counteract work satisfaction and wellbeing, then a quantitative study could pick out certain factors and see if they are correlated with more or less work satisfaction and wellbeing, compared to a control. To even make that selection, however, one must already have an idea of which factors are causally relevant. How would one know this without having some understanding of the phenomenon of work satisfaction that is not generated statistically? In this sense, qualitative and quantitative research seem to ask different types of questions and have different strengths that might complement each other.

Qualitative and quantitative approaches come with their own standards and aims. They also relate to different research traditions and corresponding philosophical biases about relativism, constructivism, empiricism, and positivism. Before we end this chapter, therefore, it is worth considering how such biases can lead to different scientific ideals, or even influence what one is willing to describe as science. In chapter 1 we defined science in the broad, continental way, as 'Wissenschaft', which from German translates directly to 'knowledge craft'. The result of having a broad definition of science, is that also the human sciences are included in the concept. In English, however, the term science is sometimes used exclusively for the natural sciences. This would mean that research from human, social, and behavioural science is excluded, and doesn't really count as science. Another distinction that is sometimes used is between the 'hard' and 'soft' sciences. Physics, chemistry, biology, astronomy, and mathematics are typical examples of so-called hard sciences, while economics, sociology, psychology, anthropology, history, and political science often fall under the soft sciences category. While the hard sciences study natural and physical phenomena that are suitable for observation, quantification, and testing, the soft sciences study human and meaningful phenomena that require interpretation. Often, therefore, researchers feel a pressure to harden the soft sciences, for instance by translating meaningful or abstract phenomena into something observable, or at least quantifiable. In Chapter 1 we explained how this is the process of operationalisation.

Returning to the imagined study of work satisfaction and wellbeing of employees. This cannot simply be observed but relies of self-reporting where the employees share their thoughts, emotions, or attitudes about their work situation. One might perform in-depth interviews with each employee, or a series of interviews, or they could answer a questionnaire. But if one wanted to operationalise something so subjective as emotions and attitudes, there are easier ways to do this. Commonly, the questionnaire will require that answers are given as a number, on a scale, say, from 1 to 5, or from 1 to 10. Then one can find out whether the person has a certain type of emotion to a very low or a very high degree. This type of survey will generate data that can be analysed and compared, for instance between workplaces or at different times, such as before and after some measures have been taken to improve work satisfaction and wellbeing. This might still be described as too soft for the natural sciences, but at least the human sciences can then use scientifically accepted methods and tools for collecting, analysing, and comparing the data.

A philosopher who has challenged the idea that the human sciences should try to imitate the methodology of the natural sciences, is Rögnvaldur 'Valdi' Ingthorsson. In 'The natural vs. human sciences: Myth, methodology, and ontology', he argues that the difference in methodology is a rational consequence of a difference in the phenomena that are studied: unconscious physical matter on the one hand, and meaningful phenomena on the other. It is a myth, therefore, that the human sciences are methodologically inferior to the natural sciences, or that they should see the natural sciences as a role model for all research. Ingthorsson also rejects the idea that only the natural sciences study objectively real, mind-independent phenomena while the human sciences do not. He argues that 'objective reality' should not be understood as 'that which exists independently of minds'. Instead, all science involves the minds of the researchers to do good quality research, design a study, formulate a hypothesis, or even interpret the data correctly. Nor should 'objective' be taken to mean 'measurable', 'observable', or even 'unbiased', while everything that is not will be 'subjective' or 'constructed'. Even socially constructed phenomena can be objectively real, he notes, such as legal systems and nations. We can see already from this brief summary of Ingthorsson's paper that Ingthorsson rejects many of the positivist biases that are common among scientists, and lead to a devaluation of the human sciences. According to him, the only definite difference between the human and the natural sciences is that they study different types of phenomena.

> In fact, the stubborn insistence that there is only one way to do science, notably the 'exact' way, may have delayed the methodological development of the human sciences... Overall the similarities are greater than the differences, especially in so far as they both deploy the same kind of rational scrutiny of the validity and reliability of its methods with respect to their subject matter, of the manner of reasoning about the data, and of the conclusions drawn on the basis of that reasoning. They judge the validity of conclusions in terms of relevance, reliability and generalizability, except that the criterions for each of these, and the degree to which they can be established, depend on the nature of the subject matter. The human sciences are, in Susan Haack's words, 'the same, only different'. (Ingthorsson, 2013, p. 41)

Ingthorsson here refers to the paper 'The same, only different', where Susan Haack argues against the idea that there should only be one accepted scientific method. Haack also denies what she sees as a false contrast between natural and social sciences, where the first seeks *explanation*, while the second seeks *understanding*. 'A letter is found that seems to show that Marilyn Monroe blackmailed President Kennedy; but the address includes a zip code, and the letter is typed using correction ribbon, when neither existed at the time the letter is dated. A meteorite is found that may show that there was once bacterial life on Mars; but what look like fossilized bacteria droppings may be artifacts of the instrumentation, or perhaps were picked up while the meteorite was in Antarctica' (Haack, 2002, p. 21). What these examples demonstrate, she says, is that both the human and the natural sciences require that the evidence is interpreted. In that respect, they both deal with meaningful phenomena. In our context here, it means that observations always require meaningful interpretation, no matter the scientific domain.

Chapter Summary

We have looked at some philosophical biases about observation and the role of the researcher as an observer. We have looked at the position that sees scientific truths as entirely empirical, neutral, and uncovered, and one position that sees scientific truths as constructed by us. Positivism fits the first position, dictating that the scientist adds nothing of themselves to the observation, but as far as possible sticks to empirical facts and data. Constructivism opposes the idea that science deals with raw or uninterpreted data, which also challenges an ideal of data centric science according to which data can be treated as independent of research agenda and context. We have also explained how the constructivist and positivist biases influence different scientific ideals and research practices. Whether these perspectives are seen as conflicting or complementary will depend on where one belongs on the spectrum between the two positions. Learning more about how research traditions are influenced by epistemological biases might anyway help scientists understand that diverging practices and standards can be equally rational and even scientific, given certain basic assumptions about observation.

Further Introductory Reading

If you are interested in learning more about relativism in the context of science, a good place to start is the short book by Martin Kusch (2020), *Relativism in the Philosophy of Science*. For more on the discussion of raw data, we recommend Sabina Leonelli's (2019) brief and accessible paper, 'Philosophy of biology: The challenges of big data biology', as well as her 2016 book, *Data-Centric Biology: A Philosophical Study*. A classic text on observation in science, is Norwood Russell Hanson's (1958) opening chapter of *Patterns of Discovery an Inquiry into the Conceptual Foundations of Science*.

Further Advanced Reading

For further reading in the distinction between the natural and human sciences, we recommend Susan Haack's (2002) paper, 'The same, only different', and 'The natural vs. human sciences: Myth, methodology, and ontology' by Rögnvaldur D. Ingthorsson (2013).

Free Internet Resources

Philosophy at the University of Edinburgh has a free online course, 'Science and philosophy', which included five lectures on relativism by Martin Kusch (2017).

Study Questions

1. What do you think about Lie's statement that the bias-free ideal of science is itself a philosophical bias?
2. What is your own perspective on the ideal of raw, uninterpreted data?
3. On the spectrum between positivism and constructivism, where would you place yourself and why?
4. Do you think that some sciences or disciplines fall more naturally into one of the philosophical biases discussed in this chapter? Explain.
5. In light of the philosophical perspectives presented here, what is your own preferred view on observation (e.g., as objective, neutral, constructed, or relative)?
6. What does Hanson mean by theory-laden observation, you think? What is his criticism of raw data?
7. If what we see is dependent on what we have learned to see, as Hanson argues, how could this influence scientific results and insights?
8. What does Leonelli say about data and data-centric science? What does she say about data travel?
9. What do you think about the separation between the natural and social/human sciences? Are your views in line with Ingthorsson and Haack or not?
10. Is it important for scientists to reflect upon the relationship between data and theory? Why, or why not?

Sample Essay Questions

1. Present and compare two or more positions on the spectrum between positivism and constructivism. Explain where you would place your own view and why. Show how assumptions related to empiricism and relativism affect scientific ideals and practices. Discuss how this might represent a barrier for collaboration across natural and social sciences.
2. One view in philosophy of science is illustrated in Fig. 6.3: that there is one reality, but that one needs plural perspectives to get knowledge about it. Discuss the ideal of scientific objectivity in light of this view, using one or more of the views presented in this chapter.
3. Present one or more philosophical views on the relationship between data and theory in science as presented in this and previous chapters. Discuss the role of data, and need for raw data, in science. Make sure you include your own perspectives in the discussion, and use examples or cases from your own education, if possible. Reflect upon which philosophical biases that could motivate more emphasis on one or the other.

References

Haack, S. (2002). The same, only different. *Journal of Aesthetic Education, 36*, 34–39.

Hanson, N. R. (1958). *Patterns of discovery an inquiry into the conceptual foundations of science.* Cambridge University Press.

Ingthorsson, R. D. (2013). The natural vs. human sciences: Myth, methodology, and ontology. *Discusiones Filosóficas, 14*, 25–41.

Kusch, M. (2017). *'Relativism', 5 short online lectures.* University of Edinburgh. https://youtube.com/playlist?list=PLKuMaHOvHA4o6JF3O-8mBeUjsT0Hv1DIY

Kusch, M. (2020). *Relativism in the philosophy of science.* Cambridge Elements Philosophy of Science Series.

Leonelli, S. (2016). *Data-centric biology: A philosophical study.* University of Chicago Press.

Leonelli, S. (2019). Philosophy of biology: The challenges of big data biology. *eLife, 8*, e47381.

Lie, S. A. N. (2017). *Philosophy of nature: Rethinking naturalness.* Routledge.

Chapter 7
Understanding and Analysing Complexity. Bias about Processes and Things

Two Ontologies, or Worldviews

Ancient Greek philosophers, also before Socrates, were concerned with ontology. Before the natural sciences developed, there was philosophy of nature, or 'physis', where philosophers speculated about the origin, constituents, and first principles of the physical reality. *What is the most foundational principle of all things? What unites everything that exists?* When we talk about ontology, therefore, we usually think of the universal, eternal, and unchanging principles of reality: that which never changes across time and space. There is, however, a contemporary debate in philosophy that can be traced back to the pre-Socratics: Could the most universal and unchanging principle of reality be that *everything is in constant change*? Perhaps change is the only constant?

We have seen in the two previous chapters how our philosophical starting point influences scientific theories, methods, and practices. Next, we will look at the ways we understand and analyse complexity, both in philosophy and science. Which basic implicit assumptions motivate our views on complexity? One might think that complexity is just what it is, and that there's only one correct way to define it. In philosophy, however, there are at least two alternative positions, motivated by different ontologies, or worldviews. In one view, complexity is understood as something that is composed of parts, where these parts have a primary and independent identity. In the other view, complexity is seen as an emergent and dynamic phenomenon in which parts cannot easily be identified or separated.

We will begin by introducing two views in philosophy: substance ontology and process ontology. These worldviews seem to represent opposing perspectives on what the most basic form of existence is: static substances or dynamic processes? Substances are things with clear identities, boundaries, or causal powers. We can easily separate one substance from another and count them, a bit like pearls on a string. Processes, on the other hand, do not have clear identities or boundaries, and cannot easily be separated or counted. How do we count wind, for instance, or fire?

R. L. Anjum and E. Rocca, *Philosophy of Science*, Palgrave Philosophy Today,
https://doi.org/10.1007/978-3-031-56049-1_7

Which of these two is taken as the correct one—substances or processes—typically depends on one's disciplinary background and education. As already discussed in previous chapters, ontological assumptions influence science and research in several ways.

First, we will have a closer look at substance ontology. This is the view that takes things and essences as fundamental and primary features of reality and tends to treat change as something that can be analysed into changeless parts. After this, we introduce process ontology. Here we will see that change, dynamism, and flux are taken as the default types of existence, which means that any form of stability must be explained, for instance as produced by multiple ongoing processes. Substance and process ontology are related to several other philosophical biases, and it is useful to also know about these and how they affect our views on complexity in philosophy and science. Note that many of the ontological assumptions that we present here can be mixed and matched in various ways, as will be explained later in the chapter. We will start by presenting two positions that contrast most clearly.

Substance Ontology: Reality Consists of Static Things, or Substances

Ontology, we saw, is typically aimed at finding some unchanging principles behind the messy reality. Everything around us is in constant change, and what's true in one moment, is not in the next. It rains now, but it didn't five minutes ago, and perhaps it doesn't rain at all in the next town or in ten minutes. These types of truths are short-lasting, and they are true only for a particular time and place. Often such truths are also highly conditional upon contextual factors. Had the conditions been different, the situation would be different too. That it sometimes rains and sometimes doesn't, is thus not the type of truth that has much scientific value. A law of nature, on the other hand, would be universally true and it would be true always, even before it was discovered scientifically. The Darwinian theory of evolution is a perfect example of a universal and unchanging scientific principle that can explain biological processes and change. Another example is Newton's theory of gravitation, which can explain the natural motion of practically any object in the universe. This is the type of knowledge that is useful for science, because it holds across all times and places and never changes.

This scientific ideal can be traced back to philosophers such as Parmenides and Plato, who thought that we can have true knowledge only of what does not change. They were motivated by a problem of change, formulated by Parmenides: *What is, cannot come to be (since it already is), while nothing can come to be from what is not. Things are what they are, but they also seem to be on their way to become something else. What they are now, they used not to be.* A world of change is to Parmenides a combination of being and non-being, of what exists and what doesn't exist. This, he thought, is not possible to comprehend. One response to this is to simply deny the possibility of change, which Parmenides did. How can something change and still keep its identity? When things change, they seem to lose their identity and become something else instead. When a nut becomes a tree, is this a nut developing into a tree, or does the nut lose or change its essence? Are we even talking about the same kind of thing?

This is how change came to be presented as a problem for ontology, and perhaps also a problem for science. The world is in constant change, which we all know. But ontology is about seeing beyond all this mess and extracting some universal and eternal truths that can explain the changing reality. This might require that we abstract some universal principles from the messy reality, such as the laws of nature that can explain what simply appears to be messy on the surface. Substance ontology is the basic assumption that, although everything in nature seems to change, all change can in principle be explained by things that do not change.

Plato's philosophy sides with substance ontology, because of his problem with change. Since the material world is in constant change, we can have no true knowledge of it. What we can have knowledge of, however, is the abstract and idealised reality that is not accessible to us through our senses, but only through reasoning. Think of the law of gravitational attraction. We cannot see the law with our own eyes. It cannot be observed directly by studying the way objects behave. Indeed, they might not ever behave exactly the way that the law predicts. Instead, we must abstract away from what we can see, and perhaps even use a theoretical model to illustrate the idea. This is primarily a rational process, in which the empirical data are explained in terms of abstract theories describing ideal objects and conditions.

Recall from Chapter 1 that Plato was mainly interested in universal, abstract existences that he called 'forms'. Think of concepts and mathematical objects. These are not particular, physical things in the world, but what we might call universals. Horse as a concept is not the same as Lukas, the neighbour's horse who is 3 years old and brown with a black tail. The geometrical form of a circle is not the particular circle that we can see when looking into a teacup. This circle has a particular size and it will only last for as long as the cup lasts. When asking: *What is a circle? What is a horse?* Plato is asking about the essence, or true being, of these types of things. These essences should not change throughout a thing's existence, and they are necessary for that thing being what it is. Essences can thus help us classify reality. What something is can be found by identifying which class it belongs to, where what is shared is the essence. All horses share the essence of being a horse, whatever that is. Science deals with essences too. The essence of water is H_2O, one might say. A cat will never be a dog, no matter how similar they can look or behave, so there is perhaps a cat-essence and a dog-essence. Perhaps DNA can represent biological essences, both of individuals and of species? Many scientists seem to think in this way.

One well-known version of substance ontology is called atomism, which is also a form of essentialism. Democritus has been referred to as the first atomist, together with his teacher Leucippus. They argued that everything in nature is composed of infinitely many tiny, indivisible, and invisible particles of various shapes. Atomism is the philosophical idea that what truly exits are these smallest static things, of substances. They are, in a sense, more real than the complex things they compose. Atomism suggests that everything, no matter how complex, is ultimately made up of some smallest parts that can compose, de-compose, and then re-compose, all without losing their identity, just like Lego bricks. The individual Lego bricks might have different shapes and colours, and these don't change when the bricks are combined with other bricks. Atomism is an extreme version of substance ontology, assuming that there are some basic units of reality that remain the same across contexts and change. Nevertheless, one might recognise the atomist philosophical bias in many scientific theories, models, or methods.

Atomism is a theory about part-whole relationships, what in philosophy is called 'mereology'. In this Lego-brick worldview, complexity consists in *mereological composition*. Parts compose in an additive way, which means that a complex whole is nothing more than the sum of its parts, primarily because those parts keep their identity and essence throughout the existence of the whole. Machines, for instance, are seen as complex because they are composed of multiple indivisible parts. These parts remain the same within the complex whole, which means that they can be assembled, disassembled, and reassembled, all without changing their essential properties. Applied to living systems, one could think of societies as composed by individuals, individual people as composed of cells, and individual cells as composed of atoms and molecules. Scientifically, it would then make sense to identify and study the individual parts in order to understand the complex whole. To think about the world as a complex machinery is the philosophical view of *mechanism*.

According to philosophers of science Daniel Nicholson and John Dupré, substance ontology survived the scientific revolution much thanks to the revival of atomism by Robert Boyle, Isaac Newton, and others: 'The atoms of early modern science were eternal and permanent in their intrinsic properties. Changes experienced in our macroscopic world were attributed to the motions of, and rearrangements of the relations between, underlying atoms, which remained unchanged throughout such interactions' (Nicholson & Dupré, 2018, p. 6). Another important influence on substance ontology is René Descartes, who promoted a mechanistic view of biology, and introduced the analytic-synthetic scientific method. This method says that when studying a complex matter, one should analyse it into its components, investigate the components individually, and then synthesise them. If a mechanical clock stops working, the way to fix the problem is to take it apart and investigate the individual parts and to repair or replace the broken ones. This is the analysis. Afterwards, one puts the parts back together, which is the synthesis. The comparison of biology and machines has led to a number of important innovations, for instance within technology and engineering, and the analytic-synthetic method is still an important tool in many sciences. It shouldn't surprise us, therefore, that substance thinking is a common philosophical bias in science.

Process Ontology: 'Change Is the Only Constant'

Although the majority of philosophers historically sides with the substance view, there is an ancient tradition for seeing change as one of the most basic ontological principles. From this philosophical perspective, any static or non-changing existence is the result of multiple dynamic processes. A process view thus denies that reality is composed of things, or substances. Process ontology also denies that there are unchanging parts or essences in nature. We are the ones who impose essences on the world, in order to classify it into neat categories. These categories, a process thinker might say, don't represent reality but rather our own substance biases.

John Dupré is a process philosopher who has spent much of his career arguing against essences in biology. In *The Disorder of Things*, he argues that our classifications of the living world are artificially clear-cut. Species, for instance, are not as easily defined as one has thought. According to Dupré, there are no biological essences or classes in the world, even though biological concepts, models, and theories rely heavily on them. Instead, one might think of all living organisms as processes. Biological organisms, including humans, only exist for as long as they are in a constant process of change. Any living organism must be in constant change or die. They also need to be in the right type of environment for healthy development and survival. This means that any apparently stable biological unity is kept so by multiple ongoing dynamic processes. It might seem like the world consists of separate things that exist independently of each other, with clear boundaries. But if we look at the living world, there are no proper boundaries between a biological unit and its environment. Rather, on the micro-level, there are continuous complex ecological interactions, such as symbiosis. All things, or substances, are then nothing but abstractions from ongoing processes in constant change, like snapshots from a movie. What appears to us as static biological substances would on this view just be a very slow process of change. Dupré argues together with Stephan Guttinger (2016) that even a virus is not a stable substance with a clear identity and boundaries, but an active ongoing cycle, or process. They explain that scientists have found that some viruses can take part in symbiotic collaborations with their hosts, which means that they can be natural parts of biological systems and processes.

Also process ontology can be traced back to one of the pre-Socratic philosophers. Heraclitus famously said that 'everything flows'. This means that if we are looking for something that is constant and unchanging, what we will find is the universal principle of change. Heraclitus is known for the saying that *no man can step into the same river twice*, meaning that the river will have changed from one moment to the next. On his view, change is the only constant.

Another philosopher concerned with explaining change, was Aristotle. He was interested in traditional ontology, and wanted to find the universal, eternal, unchanging, underlying essences of things. This he shared with Plato and many of the pre-Socratics. But like Heraclitus, Aristotle saw change as one of the basic principles of nature. All things in nature are in constant change. In addition to being, there is also becoming. Think of an acorn. An acorn exists now, but it also has an intrinsic potential to become something else: an oak tree. Once it's a tree, it's no longer an acorn. Is the tree the essence of the acorn? Or is the acorn the essence of the tree? Aristotle's ontology is an attempt to understand change, not instead of but in addition to substances. Like Plato, he was interested in essences. The essence of a thing is in its *form*, or function. All knives have a function to cut. It they cannot cut, they wouldn't be classified as knives, or at least they cannot be very good knives. Chairs might have different shapes and materials, but they share their essential function. The essence is a 'sine qua non' for belonging to a class, meaning 'without which, not'. The essence is thus a necessary, indispensable ingredient of something that makes it what it is.

Was Aristotle a process philosopher? Many would say no. Although he saw change and dynamism as basic features of nature, his interest in substances and essences has made philosophers categorise him as a substance philosopher, rather than a process philosopher. Nevertheless, Aristotle certainly stands out as one of the more process-oriented philosophers in the Western tradition, next to more recent philosophers such as Alfred North Whitehead. Contemporary philosophers promoting and developing the process thinking over substance views include, in addition to John Dupré, Johanna Seibt, Anne Sophie Meincke, Sabina Leonelli, Daniel Nicholson and Stephan Guttinger. At the moment, there seems to be a growing interest in the process ontology especially within philosophy of biology.

Also many non-philosophers are paying more interest to process thinking, and work to push their own research field away from a one-sided focus on essences, things, and classifications. In ecology and natural resource management there's an ongoing discussion about the importance of species, in particular for microorganisms such as bacteria, viruses, algae, plankton, and fungi. Some experts argue that one should focus less on classifying, mapping, and counting species and instead concentrate on their ecological functions and relationships within the system. By doing this, one could deal with whole eco-systems rather than reducing them to individual organisms and species. However, one challenge for resource management with this strategy is that it's easier for politicians and policymakers to relate to an increase or decrease of species populations than to understand which functions are vital for a healthy eco-system.

According to Dupré and colleagues, the problem of communicating and understanding processes is a result of centuries dealing primarily with substances. Models, illustrations, and concepts that we use got substances all the way down. This is why, when carrying out a large research project at Exeter on process ontology for contemporary biology, Dupré didn't only collaborate with philosophers and biologists, but also an artist, Gemma Anderson, to develop visual representations of biological processes (Anderson et al., 2019). Below is one of her artworks from the project, illustrating the protein energy landscape (Fig. 7.1).

Visual representation is essential both to the practice and the communication of science. However, whereas drawing in the past played a central role in fields such as morphology and embryology, the rise of photographic and digital technologies and the growing emphasis on molecules as opposed to whole organisms have increasingly marginalized drawing practices. Therefore, a serious problem faced in the development of a fully processual biology is

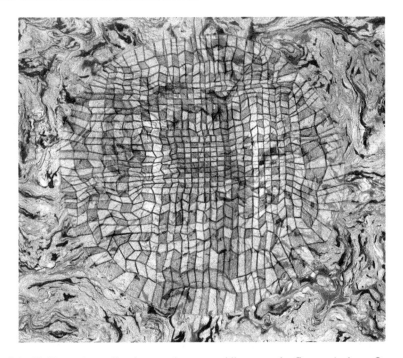

Fig. 7.1 'Fluid maze', pencil and watercolour on marbling paper, by Gemma Anderson©

that most visual representation strongly suggests a realm of static things. For example, the presentation of an organism will be of a particular developmental 'stage', typically the mature adult, which confounds the fact that this is a momentary temporal stage of the developmental process. Even where representation of something as plainly dynamic as metabolism, for example, will include arrows representing time, the natural reading will be of transitions between a fixed array of things… (Anderson, Dupré, and Wakefield, 'Representing Biology as Process', probioart.uk)

Which Is Basic: Substances, Processes, or Both?

The ultimate question concerning these two ontologies boils down to this: Do things take part in processes, or do processes produce things? It might be that static things and their interactions produce dynamic processes, which supports the substance ontology. On the other hand, perhaps what appear to be static things are just the result of dynamic micro- and macro-processes that maintain a system in relative stability. It seems that we have got three options when it comes to substance and process ontology.

One view is to say that *substances come first*. That is, one will take things, or substances, to be more basic than processes, which means that our ontology only needs to include substances. This is because substances can produce and explain processes. Processes happen precisely because there are things involved that interact. Change is perhaps a phenomenon that emerges from static substances. Or perhaps change is simply a philosophical and practical problem that requires explanation and analysis. If static things are basic, we will then need to understand how they can keep their identity and essence when undergoing change. This would include any form of change, such as when a baby eventually develops into an adult, or a seed grows into a plant. Change and dynamism then require philosophical and scientific analysis in order to be explained and understood.

The contrasting view is to say that *processes come first*. One would then assume that dynamic processes of change are primary and more fundamental in relation to static things. If this is the case, our ontology only needs to include processes, since processes both produce and explain substances. The only reason why things persist over time is that there are multiple underlying processes that keep them relatively stable. If there is an emergent phenomenon that requires explanation and analysis, it's not change, but the apparent stability of things, since everything is in constant flux. Even a cup might simply seem like a static substance on the surface, but if looking closer, one will discover that it persists only insofar as there are dynamic forces acting on and within it, keeping it stable.

There is a third possible view, which is that we need *both processes and substances* in our ontology. This view would say that processes and substances are equally real and fundamental, and that neither has ontological priority over the other. To describe reality accurately, we thus need to sometimes talk about things and sometimes talk about processes, and we shouldn't try to reduce one to the other. If we tried to find out which came first, it would end up in a chicken or egg discussion: stable things are produced by dynamic processes, and dynamic processes happen because there are stable things that interact. Philosopher Ruth Porter Groff has defended this third option, arguing in line with Brian Ellis that substances cannot be reduced to processes or vice versa. Philosophers of biology Carl Craver, Lindley Darden, and Peter Machamer (2020) also defend the view that processes and substances are equally basic, and that both are needed to understand biological mechanisms. However, other philosophers in this debate tend to prefer a strict either-or position.

We are not going to settle this ongoing philosophical debate here. Instead, we now move on to show how these different views influence science and research in various ways. Whether one sides with substances or processes, with stability or change, this will also reveal itself in scientific theories, methods, and even predictions. Our aim is to show how the two ontologies influence how science understands and deals with complexity. For this, we need to introduce some related philosophical biases that might be more familiar to scientists than substance or process ontology: reductionism, holism, and emergence.

Reductionism and Bottom-Up Causality

Scientists can be philosophically biased toward substance ontology, according to which things, or substances, are considered as more basic. To understand a process, one would then start by identifying the basic substances involved. We have already mentioned atomism, which was first a metaphysical theory in ancient philosophy, but then developed into a physical theory in the scientific revolution. For centuries, physicists have worked to identify and study the smallest physical substances: first the atom, then later the sub-atomic particles. What about biology, where everything seems to be in constant change? Are there any changeless parts there that can play the role that atoms have played in physics? Absolutely. In molecular biology, genes and proteins are seen as the unchanging drivers of changing processes.

Here is one example of substance thinking within biology. If one wants to understand how the embryo develops, then one way to do this is to start by identifying the genes and proteins involved, and then investigate how these genes and proteins interact and regulate each other's function. Any complex organism will have lots of substances and interactions, so the challenge is to identify the causal role of each substance. One standard scientific method for studying the functions of genes is to delete genes one by one from the genome in some ideal or standardised model in the lab, and then observe the consequences at the phenotype level (loss-of-function studies). To minimise complexity in such a study design, contextual factors should be kept to a minimum, and the individual tests must be identical. For this, one uses genetically identical (cloned) animal models that have been kept in isolation from various environmental influences. By using this type of approach, countless genes have been mapped for their causal role in various types of organisms. For instance, one has identified the gene for wing production in the fruit fly *Drosophila melanogaster*, and also the gene for eye production. The standard conclusion from this type of study is that the process of embryonic development is causally produced by some essential substances, namely genes and proteins, that interact in different ways in different genetic contexts. This is a form of ontological reductionism.

There are many versions of ontological reductionism, both in philosophy and in science, and we can only explain some of them here. What we described above is an example of reducing something dynamic (development) to something static (proteins and genes), but also something complex (organism) to something less complex (molecules). A reductionist will be committed to some form of ontological hierarchy, or levels in nature, where some are more fundamental or even real. One might for instance think that biological, psychological, or social phenomena are best described and explained by referring to their physical components. The physicalist version of reductionism relies on the ontological assumption that reality is ultimately physical. Physicalism counts as ontological reductionism because it says that *only what is physical can be fundamental*. This means that any non-physical facts must be reduced to fundamental facts that can then be studied by physics.

The idea of reductionism can be seen in the standard hierarchy of the sciences, where physics is the fundament (see Fig. 7.2). Above physics, we find chemistry, then

Fig. 7.2 The standard
hierarchy of science, with
physics as the foundation

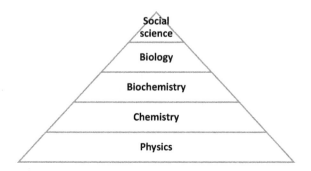

biochemistry, biology, psychology, and finally on top, social sciences. Typical for this hierarchy is that more complexity is introduced for each higher level. Societies consist of people with individual minds and brains, consisting of cells and tissues, that consist of genes, that again consist of molecules and atoms. Ultimately, a society is more complex than an individual because it includes more micro-level substances, properties, and interactions.

One version of ontological reductionism is thus the depiction of different levels of nature in a part-whole relationship, where all higher-level phenomena are composed of elements from the levels below. This is why John Dupré argues that substance ontology lends itself perfectly to reductionism, where the real 'things' and essences are found at the more basic level.

There is another version of process thinking, however, that is more reductionist, and even physicalist. Someone might argue that only physical processes are real, and that all higher-level processes must be explained by processes on the lowest level. Ontological reductionism can also be motivated by the way we think about causality. One might for instance think that reality only has one causally potent level, namely the physical one. Any non-physical phenomena would then be an effect of some lower-level causes. If so, all causal enquiries would have to start by looking for the underlying causes at a relatively lower level. This means that causal explanations, in the reductionist version, should be done bottom-up: from the lower level to the higher.

If ontological reductionism is true, then this has epistemological consequences for how we can know, understand, and explain a non-physical phenomenon. That is, we will have to look for the underlying physical causes or processes that can explain it. An example of this can be found in behavioural research, when trying to explain aversive behaviour. Researchers have observed that animals tend to move away from certain stimuli, such as some types of smell. Aversive behaviours of this type are essential for many learning programs and behavioural therapies, for instance based on negative reinforcement. Still, it's difficult to define how the higher-level phenomenon of behaviour in response to a certain stimulus should be understood. One way of studying it is by looking at the brain areas and the cell populations that are activated when a certain response is elicited by a certain stimulus. Another approach is to delete some specific biological receptor from populations of neurons and observe whether animal models with such modifications tend to lose the normal aversive behaviour to unpleasant stimuli. Based on one such experiment, aversive behaviour is explained as the activation of a certain population of dopaminergic neurons in an area of the brain called *dorsal striatum* (Xiao et al., 2020).

Some scientific theories have a clear reductionist bias, such as sociobiology and evolutionary psychology. These theories explicitly state that higher-level social and psychological phenomena can be explained using only biological principles and theories. If this is the case, we wouldn't really need sociology or psychology in addition to biology. According to Richard Dawkins, author of *The Selfish Gene*, the real agents are not us, but our genes: 'We are survival machines—robot vehicles blindly programmed to preserve the selfish molecules known as genes' (Dawkins, 1976, Preface). Another example of reductionist bias in science is when complex psychological phenomena are explained in terms of brain chemistry. In neuroscience, one seeks to detect brain activities that can be linked to higher-level phenomena such as emotions. Even love is thought to simply boil down to a certain chemical cocktail in the brain, meaning that one can produce the feeling of love by getting the chemical combination right.

Ontological reductionism also affects the health sciences, where a common idea is that that medical interventions should ideally happen bottom-up. One might then seek psychiatric diagnoses and treatments by looking at the lower level. Say a child is tremendously shy, which is typically thought of as a social phenomenon. They might be shy among strangers or in school, but not at home with the family. The child might then be diagnosed with social anxiety, which means that the shyness is reduced to a matter of individual psychology: a property of the child. If anxiety can be causally linked to certain biochemical processes in the brain, there might even be a suitable pharmaceutical treatment of it, which involves using a chemical component to alter the neurotransmissions in the brain. In this way, a higher-level social phenomenon is reduced to a lower-level chemical defect, with an equally lower-level intervention. This medical intervention then has a bottom-up causal effect: from biochemistry to social function and ability. Basically, what this form of reductionism means for health science is that we can do without social and psychological explanation and research if we have sufficient knowledge of what's going on at the basic physical level. Complexity could then be massively reduced. The psychosocial interactions between people that seem to depend on all sorts of historical, cultural, political, and personal influences, would ultimately boil down to simple stimulus–response mechanisms in the brain, cashed out in perfectly predictable input–output relationships. It should come as no surprise that this form of reductionism is a matter of controversy within the health sciences.

We have now introduced a range of philosophical biases that can motivate and support reductionism. The mechanistic worldview, atomism, and mereological composition all seem to fit well with substance ontology, bottom-up causality, as well as reductionism. All these ontological assumptions will influence scientific methodology for how to understand and analyse complexity. Given reductionism, essentialism, substance ontology, mereological composition, and bottom-up causality, the best way to deal with complexity scientifically would be to start with identifying the various parts. In any complex phenomenon, there will be multiple causal influences, and in order to find out which part plays which causal role in the whole, one will then have to consider them separately. The experimental methodology of separation and isolation is still the standard approach for establishing causality (Chapter 8), but also for evaluating risk (Chapter 9). To sum up, standard scientific methodology for dealing with complexity seems to have a default philosophical bias toward substance ontology and reductionism. We will now look at some alternatives.

Holism and Top-Down Causality

Reductionism as an ontological thesis has been challenged by philosophers, and one proposed alternative is holism, or what we prefer to call 'whole-ism'. In philosophy, ontological holism simply means that the complex whole is more than the sum of its parts. Complex wholes can consist of parts, but sometimes these parts interact with each other in ways that also influence and alter the parts themselves. If so, the parts are no longer clearly separable, and cannot easily decompose and recompose with their identities intact, unlike Lego bricks. Instead, the identity of each part is given by their causal role or interactions within the whole. This means that, outside the context of the whole, the part would no longer be that specific part. A planet, for instance, could not maintain its spherical shape or other properties if there were no external forces acting upon it. And the survival of any organism is entirely dependent on constant interacting with its environment, receiving light, heat, nutrition, and oxygen. Additionally, there is a symbiosis between an organism and various bacteria and other micro-organisms, that mutually contribute to maintain health and survival. All these phenomena could be described as holist.

Holism is an ontological assumption, but it can also be rephrased in epistemological terms, saying that one cannot predict the properties or behaviour of the complex whole from knowledge of its parts. This, however, is exactly what one wants to achieve in the scientific methodology of separation and isolation, where complexity is studied by looking at individual parts one by one. The holist would thus have to choose a different methodology to match the ontology of placing wholes first, which is a challenge, since scientific tools and methods often tend to be quite reductionist. Scientists might be in a position where their ontology is holism, but the scientific tools and approaches they must use are better suited for a reductionist ontology. From a holist starting point, a scientific investigation ought to consider the complex whole, which means to consider the contextual influences and interactions that are necessary for maintaining that whole. Rather than only zooming in by analysing the whole into parts, one would benefit from also zooming out and consider the system of which the whole itself is a part. A scientist who is a holist might not see what goes on in the lab as very representative or even relevant for what happens in real-life contexts. Instead, one might prefer to start the enquiry by studying the phenomenon in its natural context, which might be an open, complex system with countless possible influences from bottom-up and top-down. Ontological tensions between holism and reductionism thus influences epistemology: what the best method is for dealing with complexity in scientific practice.

While some research traditions will have a reductionist bias, others will have a more holist bias, and there are tensions between reductionists and holists within the same discipline. In biology, for instance, one can have holism or reductionism. From a holist perspective, one might say that it's not sufficient to study individual cells, organs, organisms, or species, but that one also needs to consider the influences and causal roles of the ecological system of which they are parts. A more reductionist approach would be to study the lower levels and search primarily for bottom-up mechanisms. Bias about holism or reductionism also influences how one thinks about the environment. While a reductionist-minded biologist might treat the environment primarily as background conditions, just as any lab-settings would be the environment for the experiment, a holist-minded scientist might see the environment as playing a crucial, top-down causal role for any biological substance or process on the molecular level.

Assumptions of bottom-up and top-down causality can offer different hypotheses and theories about the same type of phenomenon, although the reductionist would always deny the top-down alternative. Cancer, for instance, is commonly understood as happening bottom-up, from parts to whole, starting from one single cell mutation. However, some scientists have been looking at the development of cancer as a top-down process: from a cell tissue (or a society of cells) to the single cell (Sonnenschein & Soto, 1999). According to this view, cells would change their properties primarily because of a change in their environment, not their components. If cancer happens because of a failure in tissue development, rather than simply because of a genetic mutation, for instance, it could be thought of as a developmental illness. This failure in tissue development seems to provoke a change in environment, which again influences the expression of certain genes in the single cell, and thus also the single cell phenotype, changing it from regular to carcinogenic.

Let's look at another example from medicine. Attention Deficit Hyperactivity Disorder (ADHD) is diagnosed as a behavioural disorder. Mainstream medicine explains the condition as an intrinsic neurobiological disorder, with pharmacological psycho-stimulation as the main therapy. Another way to understand ADHD, however, is to see it as an effect of social and contextual factors. For instance, a public health study revealed that children born late in the year were more likely to receive an ADHD diagnosis than those born earlier in the year (Karlstad et al., 2017), suggesting that lack of school maturity in the younger children might be an important causal factor for understanding their behaviour. The question is then whether observed molecular and cellular changes in these children are the causes or the effects of their psychosocial interactions.

Note that holism can be combined with versions of both substance and process ontology. One could for instance reject atomism and say that the real substances are wholes, not the parts. A substance ontologist could also reject mereological composition and say that the whole substance is more than the sum of its parts. Process ontology could easily be holist, for instance if one assumes that the whole is produced by multiple processes where there are no 'things' that keep their identity, properties, or essence throughout the process. If so, the whole might itself be understood as an ongoing process, without any clearly distinctive boundaries. From a process perspective, this would mean that the whole is only acting or appearing as a whole because of its constant interaction with its environment.

Emergence and Demergence

We will now present a second alternative to reductionism, proposed in 'Emergence and demergence' by Rani Lill Anjum and Stephen Mumford. This is an account that involves dynamic processes, holism, and bottom-up as well as top-down causality. Going back to the Lego bricks analogy, complex wholes sometimes consist of static parts that maintain their identity, properties, and essence within that whole, and can be decomposed back to their original state. This is the idea of mereological composition, and many complex things could behave like this. What the holist would deny is that everything *must* behave this way. In living systems, the parts interact with each other and the whole, and the whole is maintained as the result of continuous and complex processes, internal and external to that whole. Ontological emergence is here the view that the complex whole has new properties and new causal powers as a result of the causal interactions of its parts, where a qualitative change also happens in the parts during this process. Anjum and Mumford call this 'the causal-transformative model of emergence'.

The transformative element also explains why one should accept ontological holism: that the whole can be more than the sum of its parts, or something else entirely. A society might be more than a collection of individuals, for instance, if no individual in separation could be thought of as having a language, culture, legal system, or economy. These are then emergent, irreducibly social phenomena. Ontological emergence in the causal-transformative sense can also be found in chemistry. Chemical bonding is often seen as a case of mereological composition, thus as an example of reductionist explanations. There is, however, a sense in which such processes could be perfect examples of ontological emergence: where each atom or molecule loses its individual properties and causal powers, and merge into a new molecule with radically different properties and causal powers. The chemical compound of water, for instance, does not simply have the properties and causal powers of its atomic components: hydrogen and oxygen. This suggests that the process of chemical bonding has altered the parts, meaning that the original parts no longer exist as they did before that composition.

> This, we say, is where we can find radical kind emergence: the coming together of the parts to form a whole involves a transformation of the parts through their interaction. Emergent powers of wholes cannot then be mere aggregates because those parts themselves change, losing at least their qualitative identity, in order to enter into that whole. And it is thus by a power entering into a relation with another that a new, holistic power emerges. (Anjum & Mumford, 2017, pp. 98–9)

Ontological emergence makes top-down causality possible, including what they call 'demergence'. Demergence is the view that the whole in the interaction with its environment can cause a change in its own parts. If so, that would count as a case of top-down causality, but one where something new 'demerges' on a lower level within that whole as part of the causal process. From a holist perspective, it seems perfectly plausible that one can influence the course of development by intervening at a relatively higher level, to produce a lower-level change. For instance, one might assume that by changing one's social situation, it is possible to causally affect one's physical properties and processes. Many social situations seem to have direct or indirect impact on our individual health, such as high blood pressure, heart conditions, or digestive disorders. Divorce, abuse, trauma, grief, poverty, unemployment, loneliness; these are all relatively higher-level situations that can make us physically ill.

> So what this tells us is that emergent powers can then act on their parts, and this is what we mean by downward causal influence. It might be useful to think of this as, to coin a phrase, demergence. Emergence is where there are new powers of wholes in virtue of causal interactions among their parts; demergence is where there are subsequent new powers of the parts in virtue of the causal action of the whole upon them. (Anjum & Mumford, 2017, p. 102)

To sum up this anti-reductionist view: assuming holism, including the possibility of top-down causality, emergence, and demergence, the causal roles of a complex whole cannot be found by analysing the parts in isolation. From this perspective, one could for instance say that the molecule of DNA has a specific causal power precisely *because* it is part of a whole cell, and of a whole organism. If extracted from the cell, DNA has no causal power and degrades in a short time. So instead of explaining the whole organism as simply caused by its DNA, bottom-up, one could say that there is a top-down causal influence from the organism in its environment to the DNA. The organism's interaction with its environment might then causally influence which genes are activated or switched off in the DNA. This requires a process perspective: Neither the DNA nor the organism would be understood as static substances with clear boundaries and identity, but instead as dynamic complex processes that include their environment.

Chapter Summary

We have explained the difference between some versions of substance ontology and process ontology. We have shown how each of these worldviews influences scientific thinking and practice. We have seen that while substance ontology is more naturally linked to essences, reductionism, mereological composition, and bottom-up causality, process ontology could motivate assumptions of holism, emergence, and top-down causality. These philosophical biases are often discipline-specific, dependent on the area of research. For some disciplines this is relatively easy to spot. For instance, emergence and holism is more common in the discipline of ecology, where one assumes that the interaction between an organism (the part) and an ecosystem (the whole) changes both. Molecular biology, however, has for a long time been substance oriented, seeking to establish the specific causal role of individual genes for the organism as a whole. For some other disciplines, however, the line might be more difficult to draw. Given the brief overview presented here, it should still be possible to reflect upon the philosophical influences on one's own disciplinary tradition, and whether it seems dominated primarily by an ontology of processes or substances.

Further Introductory Reading

For an introduction to process and substance ontology, J. R. Hustwit has written an entry on 'Process philosophy' for *Internet Encyclopedia of Philosophy*. A bit more advanced is the *Stanford Encyclopedia of Philosophy* entry on 'Process philosophy', written by Johanna Seibt. Another excellent resource is the open access book *Everything Flows. Towards a Processual Philosophy of Biology*, edited by Daniel J. Nicholson and John Dupré (2018). The book offers a range of applications to biology, but it also includes discussions of how the process ontology relate to causality, perception, and personal identity. A discussion of substance and process ontology in relation to causality and time is found in R. D. Ingthorsson's (2021) open access book *A Powerful Particulars View of Causation*, which gives a clear overview of each of the views without committing to one specific theory over the other. The relevant chapter is 'Processes and entities'.

Further Advanced Reading

Since Western history of philosophy has been dominated by the substance ontology, most texts discussing these will be written by those who are critical of substances and offer process ontology as a better alternative. It is therefore difficult to find texts that present substance ontology explicitly and favourably. A nice overview of the discussion is offered in *Process Metaphysics: An Introduction to Process Philosophy* by Nicholas Rescher (1996). For an original historical text on process ontology, see the classical work *Process and Reality* by Alfred North Whitehead (1929). There is also a collection of original texts, *Philosophers of Process*, edited by Douglas Browning and William T. Myers (1998). For further readings on emergence and on top-down causality, see *Philosophical and Scientific Perspectives on Downward Causation* edited by Michele Paolini Paoletti and Francesco Orilia (2017). This volume also includes the chapter 'Emergence and demergence' by Rani Lill Anjum and Stephen Mumford (2017). An argument for understanding viruses as processes is found in 'Viruses as living processes' by John Dupré and Stephan Guttinger (2016).

Free Internet Resources

Both the entries on process philosophy by Hustwit and Seibt are open access, in *Internet Encyclopedia of Philosophy* and *Stanford Encyclopedia of Philosophy*, respectively. At the website of Oxford University Press, the book by Nicholson and Dupré (2018) can be downloaded for free, and at the Routledge website, one can download Ingthorsson's (2021) book, also with no charge.

Study Questions

1. What characterises substance ontology?
2. How does this type of ontology influence science and research?
3. How would you describe process ontology to someone who has never heard of it? In what ways does it contrast with substance ontology?
4. Describe some ways in which process ontology can influence science and research.
5. Try to explain, in your own words, how scientific methodology can be seen as more substance oriented or more process oriented.
6. What is the difference between bottom-up and top-down causality? See if you can find examples of each.
7. What is reductionism? Can you recognise other examples than presented here of reductionist assumptions in science?
8. What are some signs of atomism, or mereological composition, in scientific theory or methods?
9. How do you understand holism, or whole-ism, and how can it affect scientific methodology?
10. Are you aware of discussions in your own discipline concerning any of the ontological views presented in this chapter: processes, substances, reductionism, mereological composition, holism, emergence, demergence?
11. How could it benefit science or education, you think, to have these discussions?

Sample Essay Questions

1. Discuss and compare how substance and/or process ontology might influence scientific theory and practice. Use examples from your own education or discipline to illustrate the ideas. Make sure you include a critical discussion of your own views and philosophical biases.
2. Present the views of reductionism, holism, and emergence. Discuss how these might relate to top-down and bottom-up causality. Explain with examples which of these biases have influenced your own education and thinking, and how.

3. Present and discuss theoretical, methodological, or practical tensions in your own discipline that might be motivated by different philosophical biases of the types presented in this chapter? How do these represent barriers for interdisciplinary collaboration? Do you see some benefits of teaching students and/or scientists to recognise and discuss implicit ontological assumptions in their discipline?

References

Anderson, G., Dupré, J., & Wakefield, J. G. (2019). Philosophy of biology: Drawing and the dynamic nature of living systems. *eLife, 8,* e46962.

Anjum, R. L., & Mumford, S. (2017). Emergence and demergence. In M. Paoletti & F. Orilia (Eds.), *Philosophical and scientific perspectives on downward causation* (pp. 92–109). Routledge.

Browning, D., & Myers, W. T. (Eds.). (1998). *Philosophers of process.* Fordham.

Dawkins, R. (1976). *The selfish gene.* Oxford University Press.

Dupré, J., & Guttinger, S. (2016). Viruses as living processes. *Studies in History and Philosophy of Science Part C: Studies in History and Philosophy of Biological and Biomedical Sciences, 59,* 109–116.

Ingthorsson, R. D. (2021). *A powerful particulars view of causation.* Routledge.

Karlstad, Ø., Furu, K., Stoltenberg, C., Håberg, S. E., & Bakken, I. J (2017). ADHD treatment and diagnosis in relation to children's birth month: Nationwide cohort study from Norway. *Scand J Public Health. 45* (4): 343–349. https://doi.org/10.1177/1403494817708080

Machamer, P., Darden, L., & Craver, C. F. (2000). Thinking about mechanisms. *Philosophy of Science, 67,* 1–25.

Nicholson, D. J., & Dupré, J. (2018). *Everything flows: Towards a processual philosophy of biology.* Oxford University Press.

Rescher, N. (1996). *Process metaphysics: An introduction to process philosophy.* SUNY Press.

Sonnenschein. C., & Soto, A. M. (1999). *The society of cells: Cancer and control of cell proliferation.* Springer Verlag.

Whitehead, A. F. (1929). *Process and reality.* The Free Press.

Xiao, X., et al. (2020). A genetically defined compartmentalized striatal direct pathway for negative reinforcement. *Cell, 183,* 211–227.

Chapter 8
Scientific Methods and Causal Evidencing. Bias about Causality

Causality in Science and Society

It's an important part of the scientific endeavour to understand causes and effects. Today we are witnessing the increasingly evident effects of unsustainable human activity: environmental pollution, shrinking glaciers, rising sea levels, extreme weather, destruction of eco-systems, loss of nature and biodiversity, conflict, war, and new diseases. Although many of these problems were caused by our scientific and technological advances, starting with the industrial revolution, it does not seem realistic to solve them without contributions from science and technology. For this, causal knowledge is vital. Finding good solutions often depends on us identifying the causes of the problem correctly and get a better understanding of the causal mechanisms leading up to the unwanted effects. We might also need to find out how to counteract further negative effects, or how to prevent or even reverse some of the destructive causal processes that have started to unfold. This is why governments rely on experts to provide them with a solid knowledge base for decision-making. So-called evidence-based policy making involves identifying a problem and the most efficient ways to remedy it. One thus needs to know how to *produce, influence, counteract, interfere,* or *prevent* certain effects, but also to identify various contextual factors that might *contribute* to the situation, positively or negatively. All these notions are causal. Without causal knowledge, how could we ever do science or make rational decisions about what to do and how to do it?

One might argue that any solution depends on understanding the complexities of a problem. Say we want to solve the problem of poverty in a population. Can we do this without first identifying the causes of poverty? Without such knowledge, how can we be sure that we are targeting the right issue? If someone is poor because they are unemployed, then finding them a job might help. If their poverty is instead caused by low income or exploitation, then lack of employment is not the main issue. This suggests that understanding causes allows us to intervene more efficiently, or at least search for ways to do so. Identifying causes is not sufficient to make a

R. L. Anjum and E. Rocca, *Philosophy of Science*, Palgrave Philosophy Today,
https://doi.org/10.1007/978-3-031-56049-1_8

decision about which intervention is best, of course. One also needs to consider the wide range of consequences that any such intervention might have. Because of this, many planning processes and public decisions require a detailed impact assessment before deciding what to do. An impact assessment is when a group of experts maps and evaluates a wide range of consequences of an intervention, looking at it from different perspectives. That way, one can reduce the risk of solving one problem, while simultaneously creating several new and perhaps much more severe problems.

Let's take an example. According to the UN climate reports written by the Inter-governmental Panel on Climate Change (IPCC), we need to replace fossil fuel with renewable energy sources to reduce CO_2 emissions and counteract further increase in global temperatures. How to find the best solutions for this in each country is a complex issue. Such a shift requires knowledge, technology, land, and financial resources. But one also needs to consider what consequences this shift will have for the local environment, the society, and the economy in order to find the optimal solution. An impact assessment therefore benefits from combining expertise from multiple disciplines, or even from trans-disciplinarity, where also local populations and other stakeholders are represented. Whether one's expertise is within renew-able energy, engineering, natural resource management, ecology, forestry, fishery, tourism, local history, or archaeology, this will influence one's perspective on which consequences to give more weight in the assessment. Since an intervention usually has some positive and some negative consequences, an impact assessment also needs to weigh and evaluate different consequences against each other.

When considering complex causal matters, interdisciplinarity is a strength but also a source of conflict. First, there are the discipline-specific perspectives, where experts are trained to have certain interests and focus while ignoring others. For instance, when assessing the impact of establishing wind power plants on land, the reports would look very different depending on the experts involved and who orders the assessment. If the report is meant to assess the impacts on birds, bats, and deer in the area, there will be other causal relationships emphasised than if one is asked to assess the impact for local businesses or infrastructure. Secondly, there is the question of how causality is best established, scientifically. As already explained in Chapter 5, there are disciplinary biases that lead to different weighing of evidence. Different disciplines tend to favour different methods and types of evidence. For instance, if there is good statistical evidence of a correlation between the intervention and an outcome, but no mechanistic evidence to support it, then an interdisciplinary group of experts might disagree over whether that amounts to sufficient causal evidence. One could then have a situation where experts from different disciplines arrive at opposite causal conclusions, based on the same pool of empirical evidence. How is that even possible? Isn't causality one single thing? Not necessarily. Looking at the diverse scientific practices and the various methods used to study and establish causality, it seems that scientists from different research traditions are searching for different things even though they use the same concept.

We will now look at different concepts of causality. For this, we must look to philosophy, where this question has been debated for more than two thousand years. While no consensus has yet been reached among philosophers, scientists seem to have already made their choice of causal concept via their choice of research methods. An efficient way to find out which concept of causality is assumed in a research tradition is thus to look at the way they study it. This is because, when accepting something as *causal evidence*, or as the *best method* for establishing causality, one first needs to have a concept of causality in mind.

We here assume that what the method is designed to find ought to match the phenomenon one is interested in studying. In Chapter 1 we mentioned the strategy of operationalism, the idea that the operation performed to empirically identify a phenomenon must be fit for purpose. Designing a method to establish causality empirically then depends on the successful operationalisation of causality. In many sciences, however, there is more than one accepted method for establishing causality. Sometimes the methods are ranked according to whether it is thought to provide weaker or stronger evidence of causality. Such ranking of evidence will indicate which concept of causality one accepts. This is not up to the individual researcher to decide, but is instead a matter of disciplinary convention. In biology, for instance, one typically ranks experiments higher than one does in economics, although this method is used in both disciplines. In medicine, large statistical studies rank higher than lab experiments, although both methods are used.

Scientists often remain unaware of which philosophical concepts of causality they have adopted via their research methods. If so, one has tacitly accepted a philosophical bias about causality. Perhaps this isn't a big problem for most empirical researchers, who might be happy to leave the discussion about the 'true nature' of causality to philosophers. Our aim in this book, however, is to show how such philosophical basic assumptions have practical implications and can become sources of expert disagreement. What one researcher sees as decisive causal evidence, the other might dismiss as poor or even irrelevant evidence. This means that two experts can look at the same empirical evidence and arrive at opposite conclusions from it regarding causality because they rank the types of evidence differently. In those cases, it could help to know about different concepts of causality and how different methods favour different concepts. It could also help to know something about the strengths and inherent blind spots of each of these concepts, and thereby of their matching methods. To provide such insights is the aim of this chapter.

Could Causality be Nothing but Correlation?

So, what exactly is causality? Surprisingly, perhaps, the greatest influence on both philosophical thinking and scientific methodology is also one of the more radical theories available: David Hume's regularity theory, proposed in his book from 1739, *A Treatise of Human Nature*. What's radical about this theory is that causality is defined as nothing more than a certain type of correlation. Hume was a strict empiricist, and only trusted knowledge that is based on sense experience. What we can observe directly about causality, according to him, is that two types of events follow each other, regularly and predictably. He used the example of two billiard balls to illustrate

Fig. 8.1 We cannot observe
the causal connection, only
A and B. Do we need more?

his idea of causality. One ball moves, touches another ball, and immediately that ball starts moving. In observing these balls, Hume couldn't find empirical evidence of anything connecting these two events. There was no 'cement of the universe', but just one thing happening after another.

While most scientists would agree with the famous saying that *correlation is not causality*, Hume proposed that a correlation is causal if it meets the following three criteria: (1) A happens before B (temporal priority), (2) A and B meets in time and space (contiguity), and (3) A and B always occur together (constant conjunction). All of these are observable features of causality. The enormous influence of Hume's theory lies not only in how it has shaped scientific methodology, but also in how almost all his philosophical opponents ended up arguing that causality must be something *more* than correlation. Causality then becomes a Hume-type correlation plus something extra. What separates the different concepts of causality is primarily what this 'something extra' should be (see Fig. 8.1). According to Hume's account, what we cannot observe is any necessary, physical, or law-like connection between A and B. This suggests that we must give up on looking for further evidence or explanation of *how* B follows from A. It is enough to know *that* it does.

We might object against Hume that, surely, there must be some law of nature or physical connection between cause and effect! What about the Newtonian forces, for instance, that clearly govern the behaviour of the billiard balls? To Hume, such laws of nature would be as speculative and unobservable as the causal connection itself. We should stick to talk about regularities, he argues, and avoid talk of laws. Hume saw repetition as vital for establishing causal regularities. It's not enough that A and B are observed together only a few times. B needs to follow A every time, at least when the conditions are the same. This way, laws of nature could be more accurately translated into claims about perfect regularities, without referring to theoretical speculations about unobservable elements.

Already here we can detect some methodological implication of Hume's regularity theory. If causality is nothing but a form of correlation, then we need to use methods that are suitable for establishing correlations. This suggests that we ought to use quantitative methods and very big data sets. According to the regularity theory, if we ever had complete correlation data, we could find all the relevant correlations in the world, and thereby all the causal connections. This might be unachievable in practice. Nevertheless, it is a scientific ideal which is increasingly widespread in empirical research and that has motivated many of the scientific methods that are used today. If it works, it means that causality doesn't require theoretical understanding, but can make do with knowledge of regular behaviours. The upside of this is that we avoid saying more than what we have empirical evidence of, such as theoretical speculations about why things behave in the way they do.

Scientists who accept Hume's regularity concept of causality tend to trust empirical data over theory (a bias we discussed in Chapter 6). Still, they might think that Hume was a bit too extreme in his empiricism. He went as far as to suggest systematic book burning to rid the world of speculative science and metaphysics (Fig. 8.2) and keep only those books that contain empirical facts or quantitative matters. In the following quote we can see where the positivists got their inspiration.

> If we take in our hand any volume; of divinity or school metaphysics, for instance; let us ask, Does it contain any abstract reasoning concerning quantity or number? No. Does it contain any experimental reasoning concerning matter of fact and existence? No. Commit it then to the flames: for it can contain nothing but sophistry and illusion. (Hume, 1748, *An Enquiry Concerning Human Understanding*, sect. 12, pt. 3)

Even if one doesn't endorse book burning, there are still many researchers who would claim that theoretical speculation has no place in science and should be avoided. We see this in the emerging trend to replace theory with large data sets and rely on so-called big data science. This leads to a tension within some disciplines, where some researchers take data and statistical tools to be sufficient for establishing causality, while others would say that data depend on theory to be causally relevant

Fig. 8.2 If we took Hume seriously... Illustration by Sheedvash Shahnia©

and informative. We will now look at a range of concepts of causality from philosophy and show how each of them would favour certain methodological approaches. Note that many of these concepts are easily combined to formulate more complex notions of causality. For our purpose here, however, it is easier to look at them separately, and concentrate on some core concepts of causality.

One Concept, Many Different Meanings

We have already said that there are many concepts of causality available, and that different disciplines and research traditions might mean very different things when they use the term. That is the reason why one disagrees over what counts as causal evidence, or at least, what counts as strong or weak causal evidence. As we go through the different notions of causality, we will also indicate which methods seem more appropriate to establish it and what forms of evidence are needed. Since there are different versions of each of these concepts of causality, we present them without attribution to specific philosophers. References to further readings within each theory can be found at the end of the chapter.

Causality as Perfect Regularity

The concept. This view is inspired by Hume's concept of causality, as we explained above, which can be summarised as follows: *Causality is a pure regularity, or correlation, where cause and effect always occur together*. In other words, whenever the cause happens, the effect follows. This amounts to a perfect regularity between cause and effect, at least when the exact same conditions are repeated. This is all there is to causality, and no further element than the constant regularity of A and B is required for a causal relationship between them.

An example. We know that excess nutrients in a lake is regularly followed by something called *eutrophication*, which is when the water turns green, slimy, and sometimes smelly. If we want to explain the causal connection in detail, one can do this by referring to more regularities. One might explain that nitrogen and phosphorus is observed to be followed by faster algae and plant growth, and that the layer of decayed algae and plant material blocks out sunlight and oxygen from the lake. Further relevant regularities might be toxin release, fish death, and other changes in the lake's ecosystem. Observations of regularities do not depend on theories of mechanisms, although the regularities could be used to develop such theories.

Advantages and limitations. An advantage of thinking of causality in terms of regularity is that it doesn't require theoretical explanation or knowledge about causal mechanisms, laws of nature, or other unobservable elements. Instead, causal knowledge can be generated by data alone. This means that causality remains an entirely empirical matter. This advantage is also its weakness since it becomes difficult to make causal claims or predictions in cases where we got no or little data available. Because repetition is key to causal knowledge, there could be no such thing as a causal matter that is yet undiscovered. Causality simply means that we have an already observed regularity. In lack of data, one might have to wait and see before making any claims about causality, which means that any request for precautionary measure would seem groundless. There can also be no unique causal event. Something that happens only once could never be a regularity. Hume admitted as much when he said that if the creation of the universe happened only once, it wasn't causal (Hume, 1748, p. 148). For this, one would have to observe the same type of event many times, under the same conditions. Even then, one cannot make predictions beyond the data, since that would amount to inductive inferences, as explained in Chapter 2. A more common criticism of regularity theory is that one cannot distinguish between a genuinely causal law and an accidental generalisation if A is always followed by B.

Methodological implications. The advantages and limitations of the regularity theory are reflected in research methods. When seeking to establish a causal correlation, one always relies on quantitative methods, such as large observation studies, big data sets, and statistical tools. If the data set is too small, or not sufficiently representative, the causal evidence is seen as lacking, or at least weak. From the perspective of regularity theory, any unresolved question of causality could be solved by access to more correlation data. A perfect data set would be the ideal situation, resulting in complete causal knowledge. Such a dream scenario is of course not realistic, but it explains the motivation behind big data science and open access to large data bases in the scientific community.

Causality as Necessary Laws

The concept. There is a long philosophical tradition for thinking of causality in terms of necessary connections. According to this concept, *a cause necessitates the effect, and guarantees that it will follow*. Causality here refers, not to the perfect regularity itself, but to what makes the regularity happen. It could be a law of nature or other types of law-like connections. This concept of causality typically involves the assumption of *causal determinism*, which means that the cause guarantees its effect. Any given set of initial conditions in the causal setup would then determine a particular effect, and if exactly the same conditions were repeated, the effect would necessarily follow.

An example. Newton's laws are the philosopher's ultimate example of causal necessity. The regularity observed on Hume's billiard ball table could then have a physical truth-maker in Newtonian forces and laws of motion. Instead of just pointing to the regular behaviour of the billiard balls, the laws of physics could explain why the observed regularities are also necessary. Causal necessitation thus provides a guarantee that also future billiard balls will behave in the same way as those that we have observed, following the same laws of nature.

Advantages and limitations. A clear advantage of the concept of causal necessity is that it makes the causal claim universal. Instead of talking only about our available data set, we can say something much more general that applies to all similar instances. That is, if the regularity between cause and effect is a matter of necessity or even determinism, then we got a rational ground for making causal predictions. Recall that Hume was sceptical of causal predictions, since we can only know what we have already observed. However, if the regularity is necessary and not accidental, we already know what will happen when the conditions are the same, even without a complete data set. The law or theory would provide everything we need to say that *all* As are necessarily followed by Bs, including future As and Bs. The disadvantage of this concept is that it depends entirely on getting the theory right. Historically, we have seen scientific theories being challenged, revised, refuted, and replaced. How, then, can we know if our current theoretical explanations are the correct and final versions? Another limitation lies in the assumption that causes necessitate their effects, *given some ideal, normal, or same conditions.* When assuming causal necessitation, then whenever B does not follow from A, it would mean that the conditions fail to meet the requirements of the theory. This type of reasoning might render the causal claim vacuously true. In other words, causal failure simply means that the conditions were not ideal, normal, or the same. How, then, could we ever falsify a causal claim? This is why we should always consider whether a law-like claim is stated as causal and empirical, or whether it instead plays the role of a theoretical stipulation or definition.

Methodological implications. When we look for causal necessity, we are seeking something like a causal law that makes the relation between A and B hold universally and without exception. Typically, this means that one needs to specify under which conditions the law applies. Many laws of physics seem to hold only under some very ideal and abstract conditions, such as the law saying that all objects in a vacuum fall with the same speed. In any normal situation, we know that feathers and cannon balls fall with very different speed, because of the air resistance. It takes a lot of physical effort to create perfect vacuum, so one might instead use a thought experiment. Other methods for producing the perfect conditions under which A guarantees B, could be use of models, simulations, or highly controlled experiments. In economics, there is little talk about laws of nature, but still one uses various models such as diagrams and equations to depict the ideal, normal, or standard situation that the theory predicts.

Causality as Necessary Conditions (Counterfactual Dependence)

The concept. When Hume proposed his regularity theory of causality, he also mentioned another concept that has become equally influential. Here, causality requires that we see beyond what actually has happened and are able to reason contrary to the fact: *If it wasn't for the cause, then the effect would not have happened. In other words, a cause must be a necessary condition for the effect.* Philosophers call this 'counterfactual dependence' because the effect counterfactually depends on the cause. There are two ways to interpret this concept: as a general or single causal claim. As a general claim, one states that *in general, A is a necessary condition for B*, while in the single claim, one states that *in this single instance, A was a necessary condition for B*.

An example. In genetics, when searching for the cause of a certain trait, one might want to find the gene or genes without which the trait cannot develop. For instance, the 'wingless gene' is so called because the fruit fly *Drosophila melanogaster* does not develop normal wings, or even no wings at all, when the gene is removed from its genome. This doesn't mean that there couldn't be other factors that causally influence how the organism develops. Still, if the trait could develop even in the absence of that gene, one might reason that it cannot have been the cause of that trait after all. This suggests that we interpret claims about an alleged necessary condition as a *general* causal claim, about all instances of that gene and that trait. As a *single* causal claim, on the other hand, we are interested only in what was a necessary condition in this specific case. For example, a young person who suffers a stroke might have had a genetic disorder disposing them for a stroke. One might then argue that, without the genetic disorder, the stroke wouldn't have happened so early in life. If so, the genetic disorder caused their stroke. This can be true even though other people could suffer a stroke also without that genetic disorder. Grammatically, we can detect the single causal claim in the retrospective expression, 'A *caused* B' (in that single case), in contrasts to the general claim: 'A *causes* B'.

Advantages and limitations. Counterfactual dependence as a general claim suggests that if B ever happens without A, then A could not be the cause of B. This is an observable feature of causality that makes it possible to separate causes from other factors that contribute to the effect without being necessary for it. One could for instance compare situations where A is present to situations where A is absent, and check whether B follows. For the single causal claim, counterfactual reasoning is more complicated but still possible. By studying a situation, one might be able to identify a significant point in the process leading up to the effect where one could say that *without* it, B wouldn't have happened. A person at a party might lean towards a loose balcony rail and fall to their death. In that case, we might reason that without that loose rail, they wouldn't have died. A problem for counterfactual dependence claims is the case of *overdetermination*: when there are two or more factors that each could have caused the effect. Say the person suffered a stroke while falling off the balcony. Then their death seemed unavoidable given either cause. Which should be correctly identified as the necessary condition for their death: the stroke or the fall? In those situations, counterfactual reasoning wouldn't work. It is false that 'if it hadn't been for the stroke, they wouldn't have died', since they would still die from the fall, and vice versa. The death was thus overdetermined, by the stroke and the fall. Herein lies the limitation. When effects are overdetermined by two or more factors, one cannot detect causes as necessary conditions. This means that any situation where there are backup systems in place to ensure that the effect happens, causality cannot be established. Biologically, there are many such backup systems in an organism, genetically and other, for vital functions. One would then need another concept of causality or find other ways to establish which specific causal factor, if any, was the necessary condition for the effect. Another limitation is that the causal chain can go on indefinitely and render any remote but necessary condition a cause. For instance, one's death counterfactually depends on one's birth, and one's birth counterfactually depends on one's grandparents' birth, and so on, all the way back to the Big Bang. Did the Big Bang thereby cause one's death, just because it was a necessary condition for it?

Methodological implications. Comparative methods with controls are well-suited for establishing counterfactual dependence and detect necessary conditions, at least for general causal claims. If in general, A is a necessary condition for B, one should not observe B when A is absent. For this one could compare two sets of correlation data. Or, if A is normally present, one might need some form of manipulation or experimentation to produce the control situation. To minimise causal complexity and avoid overdetermination, one will have to isolate A from other confounding factors. This requires some degree of idealisation or abstraction from the natural context in which A normally occurs. This could be a methodological limitation because it makes the context of study different to the context of application, and some of these differences might be causally relevant. In social science studies and in clinical medicine, one uses randomisation to make comparable groups for test and control. Randomised controlled trials is when the outcomes in two otherwise equivalent groups are compared. The reliability of this test requires that the only difference between the two groups is the presence or absence of A. Comparative studies using controls can be dependent or independent on theories of mechanisms. One might use a theory to explain why or how B causally depends on A, but causality as counterfactual dependence doesn't require any such theoretical explanation. For single causal claims, however, where A doesn't normally cause B, but did in a specific case, one might have to use mechanistic or theoretical reasoning to explain the counterfactual dependence. Otherwise, there is no way to justify that *had A not happened, B would not have happened*, since we cannot go back in time and remove A. One could, however, do an N-of-1 study, where one introduces different interventions (e.g., medical treatments), one by one, to the same situation and see which has the expected effect (e.g., recovery). The question is whether one has thereby proved that B was the effect of A and not, for instance, a cumulative effect of all those successive interventions.

Causality as Difference-Making

The concept. While counterfactual dependence requires that the effect would not have happened at all without the cause, the difference-making concept of causality is more modest: *A cause is something that makes an observable difference to the effect, compared to when it is absent.* The difference-making concept is related to counterfactual theory since one wonders what would have been different without the cause. Unlike the counterfactual concept, however, one doesn't here have to show that the cause is necessary for the effect to happen. It's sufficient to show that the cause makes an observable difference to the effect.

An example. One might test the effect on work-related accidents of a safety training program for staff by comparing companies that offer such programs with companies that don't. If there's any detectible difference in outcome, for instance a reduction in accidents at work, one might conclude that this is an effect of the safety training program. This is not the same as counterfactual dependence, which would require that there weren't any other possible ways to reduce work-related accidents, such as installing more safety equipment in the workplace or changing routines. On the difference-making concept, however, all these things could make a difference and thus count as causes.

Advantages and limitations. The difference-making concept would pick up more causal factors than any of the previous concepts. If one required perfect regularity, causal determinism, or a necessary condition, it would exclude a number of potential contributors to the effect from being causes. The difference-making concept, in contrast, could pick up on small statistical differences between two groups, or if the effect occurs in both groups but in slightly different ways, or with a time delay. For someone interested in understanding causal mechanisms or in theory development, this is an advantage, as it can work to unveil different contributors to the effect, for instance through experimentation. The feature of including more could also be a problem, if too many factors come out as causal, for instance if there is just enough observed difference to establish statistical significance in a comparative study. In those cases, causality can be a matter of controversy, since some might want the difference to be bigger in order to count as causal. Another challenge is how to identify the difference-maker in cases of causal complexity, where there are various confounding factors.

Methodological implications. As with counterfactual dependence, the difference-making concept of causality fits well with comparative methods using controls. If the difference-making concept is combined with a regularity conception, one might rely more on statistical approaches and comparisons of correlation data. This would ensure a purely empirical approach. One could also use the difference-making concept for theory development and to uncover causal mechanisms, or even laws. If so, one might use more qualitative approaches and rely less on statistical tools. For instance, one could use experimentation where more emphasis is placed on single or few experiments rather than on many repetitions.

Causality as Manipulability (Interventionism, Action)

The concept. According to this concept, *a cause is something that can be manipulated (or intervened with) in order to manipulate (or intervene with) the effect.* This is the interventionist theory of causality, and it relies heavily on the difference-making concept. The idea is that one can bring about a difference or change in the effect by acting, manipulating, or intervening on the cause. If that works, we have good indication that there is a causal link between the two, and it makes the otherwise unobservable link empirically detectible. This concept also seems to relate causality to production, suggesting that the cause somehow *produces* the effect. It also fits well with a mechanistic concept of causality, and with dispositions.

An example. One might seek to reduce CO_2 emissions of individuals, companies, or institutions by introducing carbon taxes. If the two are causally related, the expectation is that this type of manipulation will make a positive difference to emission levels. The outcome, of course, will depend on whether one can easily afford to pay the taxes and on available alternatives. Some CO_2 emissions, however, are difficult for humans to manipulate or influence, such as naturally occurring emissions from the ocean, decomposition of organic material, forest fires, volcanoes, or animals. Still, one might be able to affect some of these processes by means of counteraction or interference.

Advantages and limitations. The manipulability concept of causality is broad enough to fit well with other concepts of causality, including difference-making, regularity, counterfactual dependence, dispositions, and causal necessity. For all these concepts, one could add manipulability as an empirical test of the causal link. For instance, an observed regularity between A and B is less likely to be accidental if a change in A is followed by a change in B. One could also say that the difference-making is demonstrated best by manipulation or intervention. This is why manipulability is a suitable concept for many sciences that systematically uses intervention to establish causality. Its limitation lies in those cases where one cannot practically intervene or manipulate the situation. In social science, for instance, the causal effects of poverty, discrimination, abuse, or war cannot easily be established if they must be testable by manipulation or experimentations. The same has been said about astronomy, where one cannot manipulate the behaviour of planets. This, however, is answered by replacing the term 'manipulation' with 'intervention', where the latter could also include interventions that are not made by humans, so causality becomes less anthropocentric. One might, for instance, say that two planets can intervene on each other, or that social factors can influence each other. Still, it's worth considering whether this concept would make it impossible to establish certain causal matters even in principle. One might argue that the intervention criterion of causality is better interpreted as an in-principle possibility to counteract or produce the effect through some type of action.

Methodological implications. The manipulability concept seems a perfect match for the experimental method, whether one uses lab experiments, randomised controlled trials, simulations, or thought experiments. One might also strengthen a causal hypothesis by applying different types of manipulations, and to develop theories of mechanisms by manipulating various contextual factors. One could also use experimentation to establish different types of causal roles, such as necessary conditions, difference-makers, confounding factors, dispositions, or even causal laws, interpreted as falling under the manipulability umbrella for causal testing.

Causality as Dispositions (Causal Powers)

The concept. According to this concept, *causality happens when intrinsic causal powers—or dispositional properties—manifest themselves in an observable effect.* A disposition is a type of property that exists even before manifestation, as a real potential that can be actualised. While the manifestation is observable, the disposition might not be observable. Still, the dispositionalist will say, the dispositions itself represent real threats and promises in the world, and a real potential for causal impact. Hence some philosophers prefer the term 'causal powers' to 'dispositions'. We could say that dispositions are causally powerful properties.

An example. Examples of dispositional properties are flammability, fragility, carcinogenicity, mortality, fertility, explosiveness, and toxicity. We recognise dispositions from warning labels, where the point of the warning is not to say that their manifestation is unavoidable, but that there is a real potential for a specific effect. Dispositions are often treated as the physical truth-makers of causal regularities or counterfactual dependence. If so, the reason why smoking is correlated with lung cancer, is that cigarettes have an intrinsic disposition of carcinogenicity. Without this disposition the correlation would not be causal, according to this theory. Medicines are developed and tested for their curative dispositions, which are then manifested in combination with the person taking them. When tested against a placebo, the aim is to check that the effect is in fact caused by properties of the medicine itself and not by the patient's expectations. Dispositions can also be higher-level and belong to a whole system or a society, which means that also these can have causal powers. A society could be democratic, racist, fair, or oppressive, for instance, and education could be empowering or traumatising.

Advantages and limitations. One advantage of this concept is that one can account also for variations in effects from the same type of causal intervention, by pointing to differences in dispositional properties. Typically, one is interested in cases of where a disposition manifests itself regularity and predictably under some normal, standard, or ideal conditions. From a dispositional understanding of causality, however, it is equally important to understand these conditions better, and to see them as a vital part of causal knowledge. There might, for instance, be cases of causal failure, where the disposition fails to manifest itself when we expected that it would manifest. Rather than focusing primarily on A and B, therefore, and on whether A is regularly followed by B, one should then look at how the properties of A interact with various contexts or background conditions, and properties that might be unique for these. This also makes it possible to apply the precautionary principle, since one doesn't have to wait until causality manifests itself in observed regularities. Sometimes one wants to avoid that the effect ever happens. According to the dispositional concept of causality, knowledge of causal mechanisms would amount to an understanding of which properties could contribute, counteract, or affect the causal process and outcome. Understood dispositionally, a causal mechanism cannot be a collection of regularities or counterfactual dependence relationships but requires the involvement of real intrinsic dispositions. Since dispositional properties are often not directly observable, knowledge about them will rely heavily dependent on theoretical and mechanistic explanations. This is also why the strict empiricist is sceptical of dispositions and would not count any such causal explanations as reliable evidence.

Methodological implications. Because of the non-observable and potential nature of dispositions, there is no perfect method for establishing causality on this concept. Instead, causal evidencing requires the combination of different methods, both qualitative and quantitative, where the aim is to uncover intrinsic properties. There are many symptoms of intrinsic dispositions, including difference-making, probability-raising, and manipulability, but none of these would represent conclusive evidence on their own. Even if results from all the methods point in the same direction, causal evidence understood as intrinsic dispositions still requires the support of theories and mechanistic explanations.

Causality as Physical Process or Transference

The concept. Where most other concepts of causality are quite broad and can fit many different sciences, the physical transference concept is very specific. According to this concept, *causality is something physical—energy or similar—that is transferred from the cause to the effect.* The physical aspect of causality can be interpreted in various ways, for instance as law of nature (causal law), intrinsic dispositional property, perfect regularity, difference-making, counterfactual dependence, or manipulable mechanism.

An example. The greenhouse effect, and thus global warming, can be explained by physical processes or energy transference, including processes of radiation, energy transfer, and radiation frequencies between the sun, the Earth, and the atmosphere. Here, the causal explanation only involves purely physical elements, such as reflection of solar radiation, infrared emissions and re-emissions, and surface heating.

Advantages and limitations. The advantage of this narrow definition is that causality becomes something very specific and physically detectible. The limitation is that physical transference theory offers a highly reductionist account of causality and thereby excludes many domains where we typically talk about causality. Recall from the previous chapter that physicalism, or physical reductionism, is the idea that any higher-level phenomena and processes can ultimately be explained by referring to the physical realm. This means that causal explanations should refer only to physical causes. For instance, when looking at historical, social, or even psychological causes, one might have to refer to reductionist explanations, such as physical processes such as input, output, light particles, soundwaves, or neurons firing in the brain. Critics of reductionism argue that there are many phenomena in nature that are better explained top-down, from relatively higher-level causes to lower-level effects (see Chapter 7).

Methodological implications. As with the dispositional concept of causality, different methods are needed for establishing causality as physical transference. Relevant methods could include observations, measurements, experiments, comparative studies, and statistical methods. A causal conclusion will anyway rely on a wide range of different methods, tools, and theories about laws of physics, physical dispositions, and physical mechanisms.

Other Concepts of Causality

We have now introduced some, but far from all, concepts of causality. There are well-developed accounts of causality as *probability-raising*, for instance, stating that *a cause is something that raises the probability of its effect*. Often this is combined with a difference-making concept, because the probability-raising points to an observable difference between situations in which the cause is present and absent. Comparative methods and statistical approaches would be well-suited to pick out such probability-raisers. A further question is how probabilities are understood conceptually, which we will discuss in the next chapter.

Mechanistic concepts of causality describe *the causal link as a mechanism*, or causal *process*. For this account, it would not suffice to observe regularities: *that* the effect follows from the cause. In addition, one would need a theory of mechanism that can explain *how* the cause produces the effect. This concept can be linked to many of the theories described above, such as manipulability, difference-making, dispositions, causal laws, or physical process. A more controversial version understands a causal mechanism as nothing but the sum of multiple regularities.

Pluralist accounts of causality argue that *causality isn't a single concept, but a combination of two or more concepts*. Causality might be regularity plus counter-factual dependence, or difference-making plus a mechanism. To establish causality scientifically then requires more than one type of evidence. Conceptual pluralism might thus be an argument for methodological pluralism: that we need more than one method to establish causality. However, a pluralist account can also allow that at least one of the features of causality is established, but still assume that it would be even stronger causal evidence if all methods supported the same conclusion. The Bradford-Hill (1965) criteria of causality, for instance, pick out nine features of causality, urging that causal inquiry is a continuous process, where each new piece of evidence adds to the total pool of causal evidence.

There is a final account of causality worth mentioning, which doesn't demand that causality is something ontological that exists in the world. According to *causal constructivism, causality is a concept that is constructed by us because we find it useful to organise and interpret what we observe as causes and effects*. We use various scientific methods to pick out what we then call causality, but causality is not a feature of reality. Other accounts that are less radical, are epistemological concepts of causality, where one prefers to avoid the ontological question about what the true nature of causality is, and instead focus on the observable features of it. That way one avoids discussing whether difference-making is a symptom of a law of nature, an intrinsic disposition, a mechanism, a pure regularity, and so on. Instead, one can concentrate on identifying reliable features of causality, which is what the scientist needs to know. Ontological disputes can then be left to philosophers.

Chapter Summary

We have introduced a range of causal concepts from philosophy and looked at the methodological implications of each of them. The aim was to help clarify why causality is a source of expert disagreement over which methods are best, and how to rank and compare different methodological practices. When researchers dismiss some methods as unscientific or say they lack evidential rigour, then this reveals a philosophical bias about causality and its essential features. Being aware of the philosophical debate might lead to more critical reflection over choice of methodology, and perhaps help to see the value in diverse scientific approaches. Since all causal concepts have some advantages and some limitations, it's worth considering what exactly our chosen research methods are designed to pick out and what are their blind spots. If we accept the scope and limits of the chosen causal concept, well-aware of all the alternatives, then at least our choice of methods is an informed one. Rather than simply adopting the bias of the research tradition without reflection or awareness that such biases are involved, we can then take part in transparent methodological discourse. In cases of diverging causal evidence, this means that the scientific community can openly and constructively discuss which causal features the evidence supports, and which features have not yet been established. That way, decision-makers and other stakeholders of science can participate in the critical evaluation and discussion of causal evidence.

Further Introductory Reading

Phyllis Illari and Federica Russo's (2014) *Causality—Philosophical Theory Meets Scientific Practice* discusses a range of practical problems of causality in the natural, social, and biomedical sciences, while also giving an overview of different philosophical accounts of causality. *Causation: A User's Guide* by L. A. Paul and Ned Hall (2013) is written for both students and trained specialists and examines various contemporary theories of causality, with specific focus on counterfactual analyses. *Causation—A Very Short Introduction* by Stephen Mumford and Rani Lill Anjum (2013) offers just what the title says and is suitable for students and non-philosophers who want to familiarise themselves with the topic. If one is interested in learning more specifically about causality in science, there is the book by Anjum and Mumford (2018), *Causality in Science and the Methods of Scientific Discovery*.

Further Advanced Reading

Oxford Handbook of Causation, edited by Helen Beebee, Christopher Hitchcock, and Peter Menzies (2009), is written as a reference work for philosophy of causality and is nearly 800 pages. *Causality in the Sciences* is another wide-ranging volume edited by Phyllis Illari, Federica Russo, and Jon Williamson (2011) in which causality is discussed from the perspectives of health sciences, psychology, social science, natural science, computer science, and statistics. This book is the next level from Illari and Russo (2014) when it comes to details and technicalities, written for non-experts. There is also the *Routledge Handbook of Causality and Causal Methods* edited by Illari and Russo (2024), which presents a range of scientific problems related to causality with the aim to foster trans- and interdisciplinary reflection and discussion. There are many books that are more specific about certain concepts of causality, and here are just some of these. Counterfactual and difference-making concepts are presented in *Causation and Counterfactuals* edited by John Collins, Ned Hall, and L. A. Paul (2004). *Physical Causation* by Phil Dowe (2000) offers a version of causality as physical transference. An interventionist theory is developed in *Making Things Happen* by James Woodward (2003). The regularity view and related concepts are discussed in *Hume on Causation* by Helen Beebee (2006). *The Facts of Causation* by Hugh Mellor (2002) and *Getting Causes from Powers* by Stephen Mumford and Rani Lill Anjum (2011) present some version of dispositionalist causality. In *Hunting Causes and Using Them*, Nancy Cartwright (2007) offers a pluralist theory, and in *Causation, Evidence, and Inference*, Julian Reiss (2015) introduces a constructivist view. A concept that was not discussed in this chapter, linking causality to action and mechanism, is presented in Donald Gillies' (2018) book *Causality, Probability, and Medicine*.

Study Questions

1. What do you think about Hume's idea that there is nothing holding the world's events together, no cement of the universe?
2. How does empiricism influence how we think about causality, you think?
3. Pick the two concepts of causality that are most familiar to you. How do they fit together? Do they favour the same or different methods?
4. Which concept of causality is farthest from your own? Do you see any value in this concept at all? Explain.
5. How can different concepts of causality represent a barrier for interdisciplinary collaboration and communication?
6. What might one miss out on if one ignores other causal concepts and methods?
7. What do you think about the idea that causality is a combination of many different concepts (causal pluralism)?

8. Compare this with the idea that one needs more than one method to establish causality (methodological pluralism)?

9. Are there any methodological tensions in your discipline that might be motivated by divergent philosophical bias about causality?

Sample Essay Questions

1. Present and compare two or more concepts of causality. Discuss some strengths and limitations they might have. Use examples from your own education if you can. Explain which is your own preferred concept and consider how well it fits with the philosophical biases in your own discipline.

2. Describe one or more methods for establishing causality and identify some philosophical biases that you have learned about in this and previous chapters. How do tensions about which methods are best suited for studying and/or establishing causality relate to different philosophical biases? Use examples to illustrate or motivate the discussion.

3. Present a scientific controversy where there is disagreement about what counts as causal evidence. Analyse the controversy for different philosophical biases related to scientific evidence (e.g., best knowledge, causality, objectivity, data or theory, reductionism, holism). Feel free to include your own perspectives.

References

Anjum, R. L., & Mumford, S. (2018). What probabilistic causation should be. In *Causation in science and the methods of scientific discovery* (pp. 165–173). Oxford University Press.

Beebee, H. (2006). *Hume on causation*. Routledge.

Beebee, H., Hitchcock, C., & Menzies, P. (Eds.). (2009). *The Oxford handbook of causation*. Oxford University Press.

Bradford Hill, A. (1965). The environment and disease: Association or causation? *Proceeding of the Royal Society of Medicine, 58*, 295–300.

Cartwright, N. (2007). *Hunting causes and using them: Approaches in philosophy and economics*. Cambridge University Press.

Collins, J., Hall, N., & Paul, L. A. (Eds.). (2004). *Causation and counterfactuals*. MIT Press.

Dowe, P. (2000). *Physical causation*. Cambridge University Press.

Gillies, D. (2018). *Causality, probability, and medicine*. Routledge.

Hume, D. (1748). *An enquiry concerning human understanding* (L. Selby-Bigge, Ed.). Clarendon Press.

Illari, P., & Russo, F. (2014). *Causality: Philosophical theory meets scientific practice*. Oxford University Press.

Illari, P., & Russo, F. (Eds.). (2024). *Routledge handbook of causality and causal methods*. Routledge.

Mellor, D. H. (2002). *The facts of causation*. Routledge.

Mumford, S., & Anjum, R. L. (2011). *Getting causes from powers*. Oxford University Press.

Mumford, S., & Anjum, R. L. (2013). *Causation: A very short introduction.* Oxford University Press.

Paul, L. A., & Hall, N. (2013). *Causation: A user's guide.* Oxford University Press.

Reiss, J. (2015). *Causation, evidence, and inference.* Routledge.

Woodward, J. (2003). *Making things happen: A theory of causal explanation.* Oxford University Press.

Chapter 9
Defining and Assessing Risk. Bias about Values and Probability

Risk as a Multifaceted Notion

We all care about risk, but it's not always clear what we are talking about. Some might think of risk as a technical notion for calculating odds and guiding betting behaviour and investments. Or we might think of risk as the likelihood of some negative outcome, such as being involved in a plane crash or getting ill. When making an important life choice, we often think about how our future will look as a result of it. But how much can we know about the future? According to Hume and his empiricist criteria of knowledge, we can know nothing at all about the future. We can never know with certainty what will happen, since we haven't already experienced it. Risk, and the related notion of probability, gives us an option. Although we might not know with a 100 percent certainty what the future holds for us, we can have good reasons to have certain expectations. If we understand risk and probability correctly, we have some rational criteria for making reliable predictions.

The problem with this plan is that neither of these concepts—risk and probability—has one fixed meaning. Instead, they are tightly linked to several philosophical assumptions, both epistemological and ontological ones. First, they depend on what we think is the highest form of knowledge. For this, empiricists, rationalists, and perspectivists offer different answers (see Chapter 1). Second, they are linked to what we take as the best scientific method, which again depends on our concept of causality, as discussed in the previous chapter. A third question is how we think about predictions, as we will see.

Nevertheless, the notions of risk and probability seem unavoidable in modern science. Many scientists seem more comfortable with predicting probabilistic effects than with stating what is guaranteed to happen. This seems difficult enough, and it might be all we need to guide our decisions, both in science and in everyday life. Today, risk assessment and risk analysis are powerful and influential tools that are used beyond the scientific realm. With technological and industrial growth, we have

R. L. Anjum and E. Rocca, *Philosophy of Science*, Palgrave Philosophy Today,
https://doi.org/10.1007/978-3-031-56049-1_9

seen that technology not only offers the opportunity for economic and cultural expansion, but also represents a potential threat for human health and the environment. Risk assessment is clearly connected with the necessity of predicting these threats. Scientific risk assessment and risk analysis are today key for guiding decision-making regarding the safety and sustainability of human interventions. However, what counts as safe and sustainable will also depend on which values we hold, which interests we keep in mind, and what we see as acceptable levels of risk.

We will here look at different conceptions of risk, and the related philosophical biases that motivate them. We explain why scientific risk predictions are often diverging despite of common evidence, a phenomenon that we will call 'underdetermination of scientific risk assessment'. Finally, we explain underdetermination by identifying two types of philosophical bias about risk that influence our evaluations of it. The first type is about *moral values*, and the second is about *probability*. Without digging too much into the technical details of probability theory, we will try to explain how probability means different things in different research traditions, which then leaves us with more than one scientific method for evaluating and estimating risk.

A Very Short History of Risk

In her historical analysis of what she calls 'culture of predictions', historian of science Jamie Pietryska shows that the systematisation and routinisation of everyday forecasts had a rural origin, in the late nineteenth century in the countryside of the United States. It started because one needed reliable forecasts of the weather and of crop production, reconciling two dominant and contrasting interpretations of prediction. First, the search of order and stability, and second, the embracement of chance, uncertainty, and the fall of determinism. These two interpretations of prediction are reflected in the modern concept of risk. On the one hand, risk can be seen as a mathematical tool that allows us to control uncertainties and guide rational decision-making. On the other hand, risk can be conceptualised with its original meaning of *danger*, and thus seen as the negative side of uncertainty rather than as a tool to control it. There is, thus, a bright and a dark side of risk: control and danger.

In the early 1900s, a view was introduced that is still dominant today, which is that risk is a tool that provides control in a world of uncertainties. As such, risk is seen as an achievement of modernity and an instrument for making rational choices about an unknown future. This bright perspective on risk was introduced when the use of probability and statistics started to become prominent. Historian of science Dan Bouk (2015) identifies the start of this positivist conceptualisation of risk with the rise of the American life insurance industry. In the late 1800s in the United States, life insurance became a crucial commodity, required to get a mortgage and other services, and was increasingly offered as such to large parts or the population. To prevent bankruptcy, insurance companies needed to come up with a method to roughly predict an individuals' life expectancy. They were, in other words, in need of an objective tool to give a value to people's lives.

In this, they were helped by the powerful logic of statistics: looking at past events at population level to help predict the future risk of an individual's death. Typically, the population was divided into age groups, and for each group the average number of deaths per year was calculated. This development did not come without controversies, though. One question raised was how granular the classification should be. Clearly, the more numerous the age groups, the more individualised the predictions could be. How to determine the optimal point between individualisation ('classing') and generalisation ('smoothing')? This was not an easy question to answer. Another resistance to this statistical approach came from concerns about race discrimination. African American citizens protested that the statistics from their recent past in slavery could not give reliable predictions about their future life expectancy. This argument echoes Hume's problem of induction, stating that one cannot predict the future directly from past experience (see Chapter 2).

Despite of controversies, this positivist and empiricist view of risk grew rapidly and became popular by the 1920s. For instance, predictions about the risk of bad health outcomes at population level were now also used in public health to promote healthy behaviour of individuals. The rise of what Bouk calls 'the statistical individual' was introduced with the refinement of statistical tools, due primarily to the field of economics. It is important to notice that in this view, risk and uncertainty are treated as a dichotomy, as contrasts. Uncertainty means knowing possible negative outcomes of an action, but not knowing the odds, while risk means knowing possible negative outcomes *and* the relative odds.

This positivist idea of risk introduced some specific concepts that are still in use. In risk assessment there is an important distinction between hazard and risk. Hazard is linked to the potential to cause an undesired outcome, for instance harm to humans or the environment, and is closer to how we normally think of danger. Uranium, for instance, is more hazardous than chlorine, because the damage it can potentially provoke is bigger. Risk is a combination of hazard together with the chances of getting exposed to the hazardous substance. A ball of uranium in outer space is thus less risky than a bottle of chlorine in the hands of a baby. In order to evaluate the risk of a certain outcome to happen, one needs to calculate the chances, or probability, of exposure. This numeric, statistical concept of risk has culminated in evidence-based decision-making, in which political choices are based on scientific risk assessment. Risk analysis involves the quantification of risk into metrics such as likelihoods of undesired outcome given the exposure to some hazardous condition. Such quantification can involve the modelling of causes and effects with laboratory experiments, digital simulations, or the calculation of statistical regularities.

An example can be found in the scientific approach that started in the 1990s, to determine the risks related to long-term exposure to residues from petroleum extraction. The aim of such approach, promoted by the US Total Petroleum Hydrocarbon Criteria Working Group, was to provide standardised, scientifically sound criteria to guide the clean-up requirements of contaminated sites. For this purpose, oil was broken down into 13 fraction components, based on their molecular weight. Each fraction was analysed in a 'risk-based' manner; based on their chemical capacity for becoming bioavailable. Bioavailability was estimated by laboratory analysis of the compounds' physical properties and by modelling them together with the geological properties of the contaminated site. The result was an identification of the most volatile oil compounds that were shown to have to biggest acute toxicity: small aromatic compounds such as benzene, toluene, ethylbenzene, xylene, and light polycyclic aromatic hydrocarbons. Despite their high intrinsic toxicity, heavier compounds were considered inert, not biologically available, and therefore safe. Consequently, oil toxicity was believed to be mainly acute, and to diminish exponentially with weathering. This method of reasoning, which relies on separation, experimentation, and numerical values, was very powerful and resulted in a reduction in the requirements for cleanup in contaminated areas across the US (for details, see Checker, 2007).

This bright view of risk as a reliable support for rational and scientific choices started to be challenged already back in the 1980s. Public faith in modernity, industry, and governance began to decline as the consequence of a series of human-made disasters. Of these, the most prominent was perhaps the Chernobyl nuclear accident in 1986, in which a radioactive reactor of the Ukrainian nuclear power plant exploded and released radioactive material that spread and deposited in many parts of Europe. The field of sociology promoted the idea that humanity is facing new risk scenarios, about which we have no previous knowledge, and therefore scientific tools have limited use. While humanity has always dealt with the risk of earthquake, flood, and other natural disasters, we now face other types of risk that are not natural but created by technological development. Sociologist Antony Giddens (1999) called this new kind of dangers, produced by modernity rather than by the natural order of things, 'manufactured risk'.

The risk of long-term exposure to residues from petroleum extraction is for Giddens a manufactured risk. It's a risk for which we lack sufficient historical references to get reliable support from science and probability calculations. For example, infamous oil spills, such as the Exxon Valdez incident in 1989, where 11 million gallons of crude oil got spilled into the ocean after an oil tanker collided with a reef, showed that the scientific risk assessment approach by the US Total Petroleum Hydrocarbon Criteria Working Group had missed a big part of the picture. The spills made it possible to carry out long-term ecological studies in the polluted areas. Contrary to the US Working Group conclusions, these studies suggested that weathering of oil does indeed diminish its toxicity, but at an unpredictable pace that depends on the local environmental conditions (Peterson et al., 2003). Oil sediments can occasionally become physically trapped and therefore last longer. In addition, and in contrast to earlier predictions, they indicate that heavier compounds are bioavailable and toxic even at very low doses in developing organisms. In the long term, oil effects in the population are amplified rather than diminished through environmental interactions, for instance in the trophic chain, and therefore become substantial.

The health and environmental impacts provoked by the Exxon Valdez oil spill and other human-made disasters overshadowed the faith in the mathematical calculation of risk as a rational tool for overcoming uncertainty. On the contrary, a new wave of thought saw risk as the negative consequence of uncertainty, rather than a solution for it. Concepts such as 'ambiguity' and 'ignorance' became increasingly used and formalised, especially in connection with the development of new technologies, including biotechnology and nanotechnology. In formal terms, 'ambiguity' refers to the situations in which hazards are unknown, while 'uncertainty' refers to the cases in which the likelihood for exposure is unknown. 'Ignorance', instead, refers to the cases in which both the hazards and their likelihood to happen are unknown.

In the late 1900s, sociologists promoted the discourse on *risk society*, after the influential work of Ulrich Beck (1986). According to this view, modern society is mostly preoccupied with predicting its own technology generated risks. Not all discourse on the risk society had a negative attitude, though. Giddens himself maintained that although the tools for scientific risk assessment need to be critically considered, a society that doesn't take any risk will stop developing. We must still take some risks in order to develop and make improvements.

Risk Assessment and Values in Science

When scientists need to evaluate the risk of a certain intervention, they often generate conflicting predictions, even when a reasonable amount of evidence is collected. For instance, the causal connection between industrial practices and harmful outcomes has been extremely difficult to prove in a scientifically robust way, even when known poisonous substances are used in the production process. In many court-cases, opposing scientific judgements of risk are based on the same empirical evidence, using scientifically sound processes. When evaluating the potential impacts of contamination on human health and environment, we find ourselves facing a multifaceted problem: an overlap between conflicting socio-economic interests, conflicting scientific approaches, conflicting results, and conflicting predictions of harm. But why does this happen? Isn't scientific risk assessment supposed to provide an objective tool for analysis?

One reason for this indeterminacy is that risk evaluations don't rely exclusively on evidence and method, but also on extra-evidential premises. Evidence under determination is when evidence is not sufficient to support a scientific claim, meaning that many different claims can be supported by the same set of facts. In other words, the evidence does not determine, or necessitate, a certain conclusion. Risk assessment is according to this view not only a matter of having the best measurements and methodology, but also concerns social dynamics, value judgements, and basic philosophical assumptions. Unfortunately, these are hardly ever subject to public or governmental scrutiny.

The classical ideal of an objective science maintains that the aim of science is to provide the one truth about how things are, as discussed in Chapter 6. Contrary to this, it has in the last decades become generally accepted among philosophers of science that value-judgements are an integrative part of scientific reasoning. A question is then: What do philosophers mean by value-judgement, and which values are they talking about when they describe the scientific endeavour? One answer is that there are different types of values involved in the scientific process. The main distinction is between epistemic and non-epistemic values: those that, if pursued, help for obtaining more knowledge, and all other values, including moral ones. Epistemic values give us the norms of science and examples of such values are accuracy, simplicity, prediction capacity, and replicability. Examples of non-epistemic values are sustainability, equality, safety, and economic growth. In a famous paper by Thomas Kuhn, 'Objectivity, value judgement, and theory choice', he argues that both epistemic and non-epistemic values matter when scientists need to choose between rival theories. We will now focus on non-epistemic values and explain how they can influence the way risk is generated and evaluated by scientists.

Some philosophers have argued that non-epistemic values, including social and ethical ones, play an essential role for whether scientists accept or reject a specific hypothesis given a certain body of evidence. To understand this point, recall the problem of induction that no matter how much empirical evidence one has, it is never enough to guarantee the truth of a scientific theory or hypothesis. When we by inductive inference arrive at a general hypothesis based on a set of specific facts, there is always the risk of accepting a false hypothesis or rejecting a true hypothesis. Such risk has been called 'inductive risk'. The argument states that scientists are inclined to take the inductive risk in various degrees, and thus to make a possible mistake, depending on their social and ethical considerations. For instance, we might require more evidence to reject the hypothesis that an influenza vaccine can cause irreversible narcolepsy in small children, than to reject that the vaccine has a more trivial adverse effect, say transitory headache. In the words of Richard Rudner, 'how sure we need to be before we accept a hypothesis will depend upon how serious a mistake would be' (Rudner, 1953, p. 2).

Heather Douglas' version of the inductive risk argument is particularly powerful because it extends these considerations beyond the acceptance or rejection of a scientific hypothesis. Douglas shows how scientists must take inductive risks also when choosing a method, characterising evidence, and interpreting data. Importantly, her argument is not only descriptive but also normative. When making a technical evaluation about risk in the face of uncertain evidence, scientists *ought to* act by considering the consequences their verdict *could* generate *if they were wrong*. Let us illustrate how inductive risk influences evidence evaluation by paraphrasing one of her examples (for details, see Douglas, 2000).

In the 1990s, toxicologists were studying the toxicity of a common by-product of industrial processes, called dioxin. It was already known that dioxin could cause cancer in some tissues, but whether and to what extent the chemical is carcinogenic in the liver was still a matter of discussion. The disputed question was whether dioxin is carcinogenic at any dose, or only after a certain threshold dose. Both these intuitions were rationally justifiable given the knowledge of the time. On one hand, dioxin is indeed a poison, and it's universally accepted in toxicology that poisons are not harmful at low enough doses. On the other hand, dioxin is carcinogenic, and other carcinogenic agents such as radiations were known to be able to cause mutations, and therefore cancer, at any dose. This evaluation was important for decision-makers, since the acknowledgment of a threshold effect would imply that the chemical at low doses is not harmful and can therefore be allowed in controlled conditions.

All the scientists had access to some common evidence. For instance, exposure to dioxin was associated with tumours in rat livers at higher doses, but not at lower doses. This evidence was enough to convince some of the scientists that there is a threshold for toxicity of dioxin. After all, livers exposed to higher doses of dioxin not only had more tumours but were also more damaged. Cancer might then be a consequence of a more general liver toxicity, which is likely to be provoked by dioxin with a classical threshold effect. This evidence, on the other hand, did not convince those scientists who were already sceptical of a threshold effect. Instead, they would question the statistical power of the study. A reason why an increase rate of tumours was not detected at low doses might be that the number of animals involved was not sufficient for a statistically significant observation. From this point of view, the data of the study would not indicate a real threshold effect of dioxin, but rather a limit of detection of the test.

The key point Douglas makes with this example is the following. Since both basic intuitions—*threshold effect* or *no threshold effects*—were equally scientifically justifiable at the time, and since the consequence of adopting one over the other had an impact on the overall risk evaluation and therefore on decision-making, the choice must imply extra-scientific considerations. 'In making a choice between these positions, scientists must consider the consequences of their choice, particularly if they are wrong' (Douglas, 2000, pp. 576–7). We see that Douglas here abandons, or even reverses, the positivist ideal of an interest- and value-free science. Her inductive risk argument requires that, when making a scientific risk evaluation that has societal implications, the scientists make (and *should* make) also an explicit and active choice about which mistake they are less worried about. Therefore, non-epistemic values such as ethical and socio-political considerations are no longer seen as external, but as an integrative part of the scientific process.

Douglas gives further examples of how value choices are integrated in the scientific process. For instance, inductive risk can be involved when deciding whether a statistical result is significant. Say an observational study shows that there is a higher incidence of lymphoma among farmers who regularly handle a certain herbicide to treat their crops, than among organic farmers. How should this observation be understood? Is it an indication of a real increased risk associated with the herbicide? Or could it be just due to chance? There are conventional methods in the statistician's toolbox to help address this question. The probability that the observed difference could have occurred by chance is calculated in the form of a value called 'p-value'. More specifically, this value indicates the probability to obtain the calculated results, if there was no real difference in the incidence of cancer between the two group. Once the p-value is calculated, it can be compared to a threshold value. If the p-value is lower than the threshold of significance, then one can conclude that the observed difference between two groups is probably a real one, and not due to chance. Setting the threshold of significance is then key for the final risk evaluation. How is this done?

This is where the inductive risk comes into the picture. Conventionally, the threshold of significance can be put higher, at 0.05 (5%), or lower, at 0.01 (1%). There is no real mathematical or internal statistical reason for choosing a higher or a lower threshold of significance. In fact, these values were initially set by pure convention. Douglas argues that *if the result of the study has a societal consequence, then scientists should not apply the convention for threshold of significance blindly.* Rather, they should think whether they are most worried about getting a false positive or a false negative result and adjust the threshold of significance accordingly. Let us explain.

Consider the example above with herbicide and lymphoma. If a researcher is particularly concerned about public health and safety, the major worry might be to allow a false negative result to counsel policy. That is, one would be most afraid to wrongly interpret the observation as indicating that the herbicide is harmless, when it's in fact toxic. In this case, one should set the threshold of significance at a higher value, making it easier to get positive statistically significant results when testing toxicity. Another researcher might be more concerned with food production and with providing a certain population with an independent food system. In that case, one would worry most about producing a false positive, offering thereby a wrong interpretation which indicates that a useful herbicide is toxic, when it's in fact harmless. In this case, the threshold of significance can be set at a lower value, making it more difficult to get a positive result from a toxicity test. This way, inductive risk and value choices should be part of the scientific process, and explicitly considered, especially when dealing with matters that have a high impact on society, such as testing the risk of technologies and interventions.

This way of thinking allows for a more active role of ethics as a complement to science when carrying out scientific risk assessments. Ethics can be seen a toolbox that helps to rationally choose a good action, and to defend such a choice. Once we accept that scientists and decision-makers include value choices in the risk assessment evaluation, they must also be able to rationally make and defend such choices. From this point of view, when it comes to matters that have critical societal consequences, philosophy of science cannot be treated as separate from ethics. In 'Protection of the environment from ionising radiation: Ethical issues', scientist and philosopher Deborah Oughton (2003) explains that although scientists agree about common data, they disagree about the ethical implications. The agreed-upon data suggest that animals and plants can be damaged by radiations. The ethical disagreement is over the value that such damage should be given in the overall risk assessment and management of ionising radiations and for the evaluation of the balance of harms and benefits. Different ethical stands can be used to make the different positions explicit and rationally motivated. For instance, anthropocentric utilitarian ethics states that damage to the environment matters only when it influences human interests. In contrast, non-anthropocentric ethics can assign intrinsic value to single non-human organisms or whole ecosystems. A scientist worried about overlooking the risk of damaging individual animals will make a different inductive risk from scientists who are worried about degradation of entire animal populations, of an ecosystem, or are mainly worried about avoiding human illness. We see, then, how philosophical bias of an ethical type can deeply influence risk assessment of ionising radiations, but also of other technologies.

We have explained how the positivist ideal of a science free from values and interests has been reversed by philosophers of science in the last decades. Instead of believing that science can inform policy decisions on whether to allow new technology through an objective risk assessment, there is a move toward acknowledging that scientific risk assessment is inherently value-laden, and that any values involved in the scientific process should be made explicit and exposed to public scrutiny.

Risk and Philosophical Bias about Probability

We mentioned earlier that experts and expert committees often disagree about the significance of the available evidence when evaluating risk. Conflicting value judgements can explain some of these disagreements, but they are not the only extra-evidential assumptions at place. At the beginning of this chapter, we explained the bright and dark side of risk: that risk can be understood either with a positive connotation, as a mathematical tool to quantify and control uncertainty, or with a negative connotation, almost as a synonym for a constructed danger and uncertainty that goes hand in hand with modernity. These assumptions tend to be discipline specific. The optimistic view of risk was initiated by economists and statisticians, and the more pessimist view came from sociology. What happens, then, when risk must be studied and evaluated across different disciplines and beyond scientific contexts?

Which concept of risk we accept will likely depend on our training, but also on our other philosophical biases, for instance concerning objective knowledge, causality, reductionism, and probability. Among other things, it matters whether we accept empiricism or rationalism, bottom-up or top-down causality, and reductionism or emergence. For the rest of the chapter, we will focus on how one might think about probability. There are at least three concepts of probability, each with their own set of philosophical biases: probability understood as *frequency*, *degree of belief*, or *propensity*. An example might help to explain these. Most people accept the claim that a fair coin has a 50/50 probability of landing heads or tails if tossed, but this can mean different things depending on one's concept of probability.

Probability as Frequentism

Frequentism is one of the three theories of probability, and perhaps the most common one. The frequentist is a strict empiricist, accepting only empirical evidence as the basis of knowledge. This means that probabilities must be calculated based on data alone, without depending on theoretical speculations or causal mechanisms. For a frequentist, therefore, the only rational basis for claiming that the tossed coin has a 50/50 probability to land head or tail is that one has empirical data from previous coin tosses. Specifically, one must have counted how frequently coins land heads and tails over a very long series of trials and found that they landed tail half of the times and head half of the times. The probability of the coin to land head is then equal to the proportion of that outcome over a sequence of coin tosses, which should be half of the times for a fair coin.

Note, however, that such a probability estimate should never be made from only a few observations, since it's perfectly possible to throw ten heads in a row even with a fair coin. Still, the distribution of heads and tails should become increasingly stable toward 50/50 with more trials. This is the law of increasing stability of statistical frequencies, which is one of the fundamental postulates of the standard frequency theory, advanced by Richard von Mises in *Probability, Statistics and Truth*. In Mises' frequentist framework, probability is the mathematics of repetitive events which can have different attributes. He calls such repetitive events 'collectives'. Coin tosses are an example of collectives, where the possible attributes of each repeated event are head or tail. Within frequentism, probabilities are always attributed to collectives, not to a single event. From observing a single coin toss, all we can know is the actual outcome, and not the proportion of such outcomes over a series of similar trials. This means that it's not possible to calculate a frequentist probability for a unique case that cannot be repeated.

The frequentist conception of probability was originally postulated to study a theoretically infinite and identical repetition of collectives, which can be done in theory but not in practice. Frequentism thus presupposes that there is the possibility— at least theoretically—of an infinite number of repetitions. Also, one needs to repeat many instances of the exact same conditions. When evaluating frequentist probability, one needs to consider the following question: How similar are the collectives used to calculate the frequency of outcomes to the single event for which we want to make a prediction? If the answer is 'not very similar', the prediction becomes less relevant. For the frequentist, therefore, the more and better data one has, the more accurate and trustworthy the probability prediction based on it will be. Recall that ideally, any empiricist would want to have a perfect data set, which in the case of the coin means that one has infinitely many tosses. In practice, however, since one never has infinite time or resources to gather a perfect data set, one usually settles for one that is big enough.

Within this framework, scientists typically use statistical models and tools to calculate the relative frequency of a certain outcome in a sequence of events, using large data sets. This is the reasoning behind many risk calculators. A person's risk of getting a certain disease will in such calculators be generated based on statistical data from similar cases, which ideally should be their perfect twin population. A practical challenge for frequentism is how to best define the relevant collective, or population, from which probability should be calculated, since different collectives will generate different probability estimates. For instance, as a healthy vegetarian with high income, the risk of a heart attack might be considerably lower than as an ethnic minority with heart conditions in the family.

Probability as Degree of Belief, or Credence

Returning to our coin, we might say that the 50/50 distribution of outcomes into heads and tails is not what's important to our prediction of the coin toss. What we want to know is the actual outcome of this specific toss. Will the coin land head? Or will it land tail? By stating a 50 percent probability of the coin landing head in the next toss, what we are estimating is not the probability of the outcome itself, but our own degree of confidence in this particular outcome. Probability is then about our subjective degree of belief, or confidence, in a certain outcome, given the evidence available to us. The idea that degree of probability should mean degree of belief, was suggested already by Augustus de Morgan's laws of probability in the nineteenth century, and then refined by the independent work of Frank Ramsey and Bruno de Finetti about 100 years later (Ramsey 1926). Probability is here understood as a subjective and epistemological matter, and not as a real ontological feature of the world. Probabilities are about how confident we can be about the predicted outcome.

Within this framework of credence, the actual outcome of a coin toss is a deterministic matter. This means that the outcome will be strictly determined by the toss itself including all the initial conditions. If the coin landed heads, it was determined to do so already in the moment of the toss. Theoretically, this means that if one repeats the coin toss perfectly in all details, the exact same outcome should be guaranteed. The only problem with predicting the outcome is a practical one, since our knowledge is always limited. Only an omniscient creature, such as God, could know the actual outcome and would therefore not need to make probabilistic claims. Probability is only needed when we lack information. Ideally, we will have so much information that we can be 100 percent confident in an outcome. This shows a difference between credence theory and frequentism. While a complete data set would help the frequentist make an accurate probability claim, a complete data set is for a credence theorist the same as perfect information and certainty. In other words, there would be no need to state one's belief in probabilities if one knew enough to predict the actual outcome.

Importantly, since the probability is a measure of degree of belief, it needs to be updated every time new evidence is acquired. But how exactly should degree of belief be updated in light of new evidence? It's not sufficient to make an educated guess. Instead, there are precise rules about the way we should learn from experience, and for guiding the evaluation on how strongly a piece of evidence supports a hypothesis. Ramsey and Finetti answered by drawing a close relationship between belief and action, where action is exemplified by a person's betting behaviour. A way to measure someone's confidence in an occurrence is to measure their betting rate in an imaginary gamble, which in turn shows their willingness to bet on the occurrence. Crucially, the betting behaviour needs to be pragmatically rational and coherent. For Ramsey and Finetti, a rational agent is defined as someone who would not willingly commit to a bet with sure loss. This implies that personal beliefs are rational and coherent if and only if they satisfy the rules of probability theory.

We see that, although probability is understood as subjective degree of belief, the theory comes with a normative commitment to the rules of probability calculus, and to Thomas Bayes' theorem. The theorem is fundamental to this framework for probability and postulates how beliefs should be updated in light of new evidence. Our updated beliefs are called 'posterior probabilities'. These are estimated based on the combination of new evidence—'prior probabilities'—and the likelihood of the event that constitutes the new evidence occurring *if* our prior hypothesis is correct. According to Bayes' theorem, the probability of our hypothesis H being true, given the empirical evidence E (posterior probability or P(H|E)), is equal to the probability of seeing the evidence E whenever the hypothesis H is true, multiplied by the prior probability of H, and all divided by the probability of E:

$$P(H|E) = P(E|H) \cdot P(H) \big/ P(E)$$

For instance, let's imagine that someone has been suspecting since a couple of days that their son is getting a cold and want to evaluate such hypothesis in the light of the new evidence that he sneezed 20 times today. The probability that the hypothesis is correct given the new evidence (posterior probability), equals the probability of someone sneezing 20 times when they are getting a cold multiplied by the (prior) probability of getting a cold, divided by the probability of sneezing 20 times in a day, with or without a cold. Note that the prior probability P(H) is the probability of a hypothesis being true prior to getting to see the new evidence. Assigning such prior probabilities is not a straight-forward task, however, and different versions of Bayesianism are divided on the question of the strategy one might do it. A version of Bayesianism called objective Bayesianism provides some common criteria for how to estimate the likelihood of the prior probabilities. Still, it is important to note that Bayesianism, including so-called objective Bayesianism, sees probability as subjective credence.

Probability as Propensity

The propensity theorist sees probabilities as referring to dispositional properties. It's in virtue of its properties that the coin has a disposition to land head or tail. One can find these dispositions by studying the coin and its context. The coin has two sides, and if the weight is equally distributed, and the surface on which it will eventually land is flat and solid, then there are good theoretical and practical reasons to predict its probability to land head or tail as 50/50. One might even say that the 50/50 distribution is produced by the dispositional properties of the coin, giving the coin a causal potential, or *propensity*, to behave in that specific way. Note that from a dispositional point of view, the fact that a coin has certain properties might not guarantee a perfect distribution of actual outcomes into 50/50 heads and tails. Whether a coin is fair depends entirely on the dispositional properties of the coin and the environment in which the coin is tossed. If we remove or reduce gravity, the coin will behave differently, since the environment then has different dispositional properties. Probabilities are thus attributed to the whole setup, to predict the coin's potential behaviour within a system. In practice, this can be purely hypothetical, since the coin might never be tossed for as long as it exists, but it still has the propensity.

There are different versions of propensity theory, but one common feature is that propensities refer to a single case and gives a sort of explanatory understanding to probability. This idea of probability as propensities was first expressed around 1910 by Charles Peirce, who described the probability of a die landing 1 with a frequency of 1/6 as the 'would-be', or property, of the die. Karl Popper developed the idea and described propensities as dispositional properties of singular events in 'The propensity interpretation of probability' and in *A World of Propensities.* One can draw a distinction between *long-run* and *single-case* propensity theories. For long-run propensity theorists, such as Donald Gillies (2000), propensities are responsible for generating certain values of relative frequencies, in a series of repetitions. Gillies argues that the long-run frequency, if it exists, will define the strength of propensities. For instance, a die has a strong propensity to land on 3 with the long-run relative frequency of 1/6. This theory combines the ideas of propensities and frequencies, where the propensities might represent the ontological aspect of probability while frequentism covers the epistemological parts, of how to estimate the strength of propensities. Single-case propensity theorists, such as Karl Popper and Hugh Mellor, deny any such links with frequentism and empiricism, and place more emphasis on theories of mechanisms, dispositions, and potentiality.

In contrast to the subjectivist credence view, a propensity theorist sees probability as an ontological and objective matter, not as epistemic and subjective. This is why the propensity theorist can accept the attribution of probabilities to a unique case, contra the frequentist assumption. Ontologically, single propensities are given by the unique combination of dispositions and those dispositions' degree of tendency toward certain outcomes. Such propensities are independent of what has happened elsewhere, or in previous cases. In this framework, long-run frequencies may be indicative of a propensity of a single occurrence, but do not need to be. If one can observe a frequency of occurrence, such as the 50/50 distribution of outcomes of heads and tails in a coin toss, this frequency is produced by the propensities. In contrast, the frequentist says that frequency of occurrence determines the facts of probability.

One final issue to mention is the question of how single-case propensities should be expressed, and whether they should obey to the rules of probability calculus. Unlike the case of long-run propensities, indeed, there is no clear link between single-case propensities and frequentism. Philosophers disagree over this issue. Some maintain that the use of calculus is essential for propensities to have a practical, empirical, and testable scientific value, and therefore one should postulate that single-case propensities must follow these rules. A different take is found in our own preferred version of propensity theory (see Anjum and Mumford 2018), which favours a singular and qualitative description of propensities, rather than a purely numerical one. Since propensities are generated by dispositions, or intrinsic qualities of things, and are also determined by the disposition's magnitude or intensity, they cannot be directly derived from statistical frequencies. The propensity of a certain outcome to happen depends on the presence or absence of a specific combination of dispositions, which might be a unique case. The propensity for an outcome therefore cannot be derived from numbers and scores alone but must include theoretical and mechanistic knowledge of *how* that certain outcome could or could not happen.

While the application of frequentism and Bayesian credence to scientific risk assessment are predominant in the process of science and scientific risk assessment, not many are familiar with the propensity view. How, in practice, might a single-case propensity theorists approach risk assessment? The simple answer is: by finding out more about the local context. For this, one needs to look at the dispositional properties of the situation and linking them to the best knowledge about causal mechanisms and processes. In Fig. 9.1, the propensity of falling into the pit and hurt oneself from a jump will depend on the qualities of the environment and the person jumping. Even if there is no similar situation observed previously, we can clearly see how the different situations will have different risks. This means that single propensity theory allows precautionary reasoning, even when there are no data of harm yet available.

A criticism of the propensity theory is that it relies heavily on getting the dispositions and theories of mechanisms right. It's also not clear how one can give a precise numeric value to probability and risk, based only on theoretical and mechanistic knowledge. If probability is primarily a qualitative matter, however, as suggested by the single-case propensity theory, then this might not be a real problem. Instead, one might say that statistical frequencies tell us how often the outcome has happened, while the qualitative propensity should help us predict the probability of that outcome in this specific case.

When Life Is Not a Game…

How would these different concepts of probability play out in the context of scientific risk assessment? In real-life situations, evaluating the risk of new technologies or interventions is a complex, multidisciplinary task. On the one hand it must usually involve interdisciplinary teams, because it requires a combination of technical expertise, knowledge of statistics, and competence in evaluating various lines of evidence. On the other hand, the available empirical evidence must be weighed against ethical and social concerns.

Generally, experts might agree that the broader the evidence base is, the better the risk evaluations will be. But risk assessment is not basic research, which can take decades or even centuries. Because risk evaluation is needed to inform policy, it must be performed within some reasonable time limits. There is a delicate balance between getting enough evidence for a robust risk assessment and informing prompt actions to avoid possible harm. A common problem is therefore to decide when there is enough evidence to draw reliable conclusions about risk. The answer is usually affected by one's philosophical bias about probability, but also by one's ethical and social concerns. What are the known unknowns? Do we need to observe enough

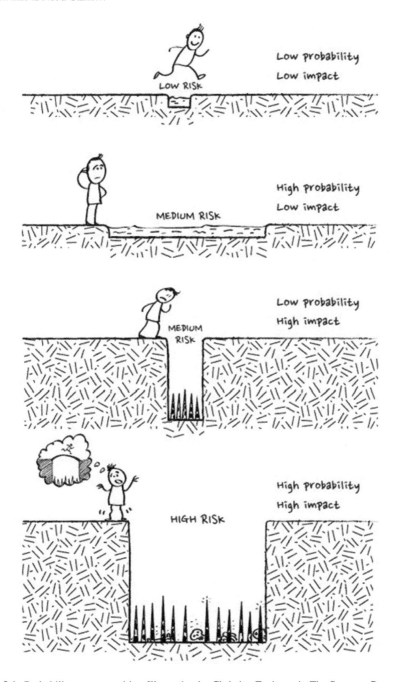

Fig. 9.1 Probability as propensities. Illustration by Christian Espinosa in The Smartest Person in the Room, 2021

repetitions? Should we rely on evidence of mechanisms underlying the risk? And what is at stake, if we go for one option or the other? Let's illustrate this with an example.

In the early 2000s, scientists were trying to understand whether living next to areas polluted by residues from petroleum extracting activities could be related to long-term health risks. This was a socially important question, since large communities in South America were exposed to this type of industrial waste. If health risks were identified, then the polluting corporations could be asked to clean up the areas. However, the scientists disagreed on whether there was enough evidence of health risk to support such demand.

One common method for investigating the matter would be a frequentist one: How many people who have been living next to the polluted regions got the long-term health effect of interest? This number could be used to estimate the risk of people living under similar conditions getting sick in the future. Of course, this is complicated by the fact that in the same areas there was a concentration of more types of pollutants than just oil residues, and there were not many healthcare and prevention measures. Even if a number of health effects were recorded in a certain number of people, how could one be sure that they were caused by the exposer of interest, in this case the oil residues? The answer for epidemiologists is that one needs large and good population studies, in which a large number of variables gets recorded and statistical models are used to distinguish confounding variables (those we are not investigating) from oil pollution.

The trouble was that the affected areas were mostly in remote regions, where demographic information at the time were scarce, health registries missing, and sickness and death under-reported. On top of this, it was difficult to get precise data about exposure to contamination. Population studies were therefore extremely difficult to conduct. What could be done? In this matter, epidemiologists disagree, and the disagreement seems motivated by different philosophical biases. On the one side, epidemiologists Anna-Karin Hurtig and Miguel San Sebastián conclude that there is sufficient evidence to recommend a clean-up of the polluted areas and the establishment of systems for cancer surveillance and environmental monitoring.

> The results suggest a relationship between cancer incidence and living in proximity to oil fields, although this ecologic study cannot lead to causal inference. However, the possibility of a causal relationship is supported by several criteria. First, the strength of the association between the outcome and the exposure. Second, there has been considerable attention devoted to the biological mechanism by which some of the components of crude oil (benzene, PAH) could increase cancer risk. Third, consistency with other investigations is apparent after reviewing the body of literature that associates oil pollutants and cancer. Fourth, by using surrogate data that are representative of several decades of oil pollution exposure, a plausible time sequence from exposure to development of disease can be inferred. Further research is necessary to determine if the observed associations do reflect an underlying causal relationship. …Meanwhile, an environmental monitoring system to assess, control and assist in elimination of sources of pollution in the area, and a surveillance system to gain knowledge of the evolution of cancer incidence and distribution in the area, are urgently recommended. (Hurtig & San Sebastián, 2002, p. 1025)

These scientists refer to different types of evidence that could support the observed statistical association between illness and exposure, such as existing knowledge about known mechanisms of harm of oil contamination, results from animal studies, and other evidence supporting the plausibility of this link. They also refer to properties observed in the local context, such as the high level of oil contamination in drinking water and heavy metal levels in the blood of residents. In 'Epidemiology of the angels', a commentary to Hurtig and San Sebastián, epidemiologist Jack Siemiatycki argues that the study is statistically too weak to make any public health recommendations. He calls their study 'a bold attempt to use imperfect data to derive scientific knowledge' (Siemiatycki, 2002, p. 1029). The evidence, he says, is too weak to support the hypothesis of a causal link between oil contamination and cancer incidents.

> Epidemiological research is sometimes used as a cover of scientific legitimacy in calling for sensible public health precautions. While this definitely puts epidemiologists 'on the side of the angels', it also risks compromising the scientific credibility of epidemiology. The paper by Hurtig and San Sebastian does not represent the most egregious example of such a tendency. But the apparent reach for suggestive results where such suggestions are at best hints, and the ease with which public health recommendations are made, suggest that the authors may have been leaning on the recommendations before the data were in and the evidence assessed. On the other hand, in the real world of lobbying and public policy, it seems that epidemiologists are sometimes 'caught between a rock and a hard place' when they try to simultaneously satisfy their rigorous scientific principles and their public health principles. (Siemiatycki, 2002, pp. 1028–9)

The disagreement is motivated by conflicting philosophical bias about causality, probability, and value. Siemiatycki seems to assume the empiricist and frequentist concept of risk, where the answer is always either to trust the data or, if data are obviously bad, to try to get better data. For instance, one could improve the health registries of the area, or perform large cohort studies following up people living in differently polluted areas for decades, ideally from birth, and record a number of variables to compare. It's of course very unfortunate, as Siemiatycki notes, that people need to stay for longer time in a situation that might be risky. But scientific risk assessment can only conclude what data allow. This suggests a bias about value, since his primary concern is to avoid false positive results from limited population data. Hurtig and San Sebastian, in contrast, seem most worried about the false negative results. They are worried that people keep living in a risky situation when the scientific data do not reveal the risk. The inductive risk argument applies in this situation because the values of the researchers influence their risk evaluation. Moreover, we see that Hurtig and San Sebastian refer to local processes, causal mechanisms, and contextual findings to compliment the weak statistical evidence. Their arguments reveal that they are interested in the propensity for toxicity to happen in this specific context.

How would a Bayesian probability analysts approach this situation? We can suppose that they would agree with Hurtig and San Sebastian that all the available evidence must be used to update the credence. Bayesians are prone to update priors with every valid observation, not only with population studies. On the other hand, Bayesians would require that every prior could be assigned a value. Qualitative findings would have to be converted into numeric values and the final numerical answer would express the level of confidence one has in the outcome (oil toxicity in this case). Such answer would have to be rigorously calculated following the rules of logic and probability calculus.

Chapter Summary

We have described some of the philosophical biases that can be involved in scientific evaluations of risk. Specifically, we have talked about value judgements, which can be seen as philosophical bias of an ethical type. We have presented inductive risk arguments and explained how such arguments demonstrate that value choices are and should be integrated in the scientific process of risk assessment. We also have discussed different basic assumptions about the nature of probability: frequentism, credence, and propensity theory. Each of these justifies a different approach to risk evaluation: statistical, Bayesian, or qualitative. We saw that all these approaches offer something that can be useful to expand our knowledge. The question is which we take to be more basic. An open question for reflection is whether making our value judgements and philosophical biases about probability more explicit can facilitate interdisciplinary risk assessment and transparency of the scientific evaluation. In the next and final part of this book, we present some further cases of scientific controversy where risk is a contended matter, among other philosophical biases.

Further Introductory Reading

Mauricio Suárez (2021) has written an excellent introduction to probability theory for the Cambridge Elements series in Philosophy of Science: *Philosophy of Probability and Statistical Modelling.* For an accessible introduction to ethical biases in risk evaluation and management, we recommend Deborah Oughton's (2003) paper 'Protection of the environment from ionising radiation: Ethical issues' (2003). If you are interested in an historical view on the rise of frequentist view on risk, you might enjoy Jamie Pietryska's (2017) *Looking Forward: Prediction and Uncertainty in Modern America*, as well as Dan Bouk's (2015) *How Our Days Became Numbered. Risk and the Rise of the Statistical Individual.* The dystopic view of risk in the modern society is described by Ulrich Beck (1992) in *Risk Society: Towards a New Modernity.* If you want a shorter read, see 'Risk and responsibility' by Anthony Giddens (1999). For an introduction to the inductive risk argument, see Heather Douglas' (2000) 'Inductive risk and values in science', which is written in an accessible way with extensive use of examples.

Further Advanced Reading

A detailed conceptual overview of risk assessment, formulated in terms of misunderstandings and pitfalls, can be found in Terje Aven's (2010) *Misconceptions of Risk*. In general, we recommend looking into Aven's work if you are interested in the technical and conceptual aspects of risk analysis and management, and the contrast between the frequentist and the Bayesian approach to this. Advanced readings about the propensity view on probability can be found in Karl Popper's (1959) 'The propensity interpretation of probability', as well as in Donald Gillies' (2000) 'Varieties of propensities'. For a detailed account of non-epistemic values in science, see Thomas Kuhn's (1977) 'Objectivity, value judgment and theory choice'. In 'From ideal to real risk: Philosophy of causation meets risk analysis' (Anjum & Rocca, 2019), we explain how different approaches to risk analysis can be seen as generated by different philosophical biases about causality.

Free Internet Resources

Stanford Encyclopedia of Philosophy has an entry on 'Interpretations of probability', which offers a nice overview. The *Science for Policy* podcast series features a podcast in which Douglas discusses her inductive risk argument. The title of the episode is 'Heather Douglas on how values shape science advice'. In our open access article, 'Why causal evidencing of risk fails. An example from oil contamination' (Rocca & Anjum, 2019), we present our own view on the philosophical bias that motivate different risk assessment in the case of toxicity from oil residues.

Study Questions

1. Try to explain, in your own word, the concept of inductive risk and the idea that values are a constitutive part of scientific risk assessment.
2. Explain the optimistic approach to risk and the way it evolved after some big environmental disaster such as the Chernobyl nuclear accident in the 1980s.
3. In your own words, explain the idea of frequentism.
4. How can one detect the influence of frequentism on scientific methodology and thinking about risk?
5. What is the credence view on probability, and how does it contrast with frequentism and propensity theory?
6. Can we say that credence fits with a deterministic view on causality and why?
7. What were the two versions of propensity theory, and which could be combined with frequentism?
8. Do you see any practical challenges for risk evaluations from propensity theory?

9. How do these interpretations of probability relate to qualitative and quantitative methodology?
10. How does bias about value affect risk evaluations and recommendations from these, you think?

Sample essay questions

1. Present the three philosophical theories of probabilities discussed in this chapter: frequentism, credence, and propensity theory. Explain how they motivate different concepts of risk. Discuss ways in which your own discipline is influenced by philosophical bias related to risk and probability.
2. Present and analyse a scientific controversy (theoretical, methodological, or practical), preferably from your own discipline, that might be motivated by different tacit philosophical biases about probability or risk.
3. Explain Douglas' view on inductive risk. Discuss critically the idea that values play an important role for scientific risk evaluation, using one or more examples. Do you agree or disagree that embracing values as integrative of scientific risk assessment can make science more democratic? Motivate your arguments.

References

Anjum, R. L., & Mumford, S. (2018). What probabilistic causation should be. In *Causation in science and the methods of scientific discovery* (pp. 165–173). Oxford University Press.

Anjum, R. L., & Rocca, E. (2019). From ideal to real risk: Philosophy of causation meets risk analysis. *Risk Analysis, 39*, 729–740.

Aven, T. (2010). *Misconceptions of risk*. Wiley.

Beck, U. (1992). *Risk society: Towards a new modernity*. Sage.

Bouk, D. (2015). *How our days became numbered: Risk and the rise of the statistical individual*. University of Chicago Press.

Checker, M. (2007). "But i know it's true": Environmental risk assessment, justice, and anthropology. *Human Organization, 66*, 112–124.

Douglas, H. (2000). Inductive risk and values in science. *Philosophy of Science, 67*, 559–579.

Giddens, A. (1999). Risk and responsibility. *The Modern Law Review, 62*, 1–10.

Hurtig, A.-K., & San Sebastián, M. (2002). Geographical differences in cancer incidence in the Amazon basin of Ecuador in relation to residence near oil fields. *International Journal of Epidemiology, 31*, 1021–1027.

Kuhn, T. (1977). *Objectivity, value judgment and theory choice*. University of Chicago Press.

Oughton, D. (2003). Protection of the environment from ionising radiation: Ethical issues. *Journal of Environmental Radioactivity, 66*, 3–18.

Peterson, C. H., Rice, S. D., Short, J. W., et al. (2003). Long-term ecosystem response to the Exxon Valdez oil spill. *Science, 302*, 2082–2086.

Pietryska, J. (2017). *Looking forward: Prediction and uncertainty in modern America*. University of Chicago Press.

Popper, K. (1959). The propensity interpretation of probability. *British Journal of Philosophy of Science, 10,* 25–42.

Ramsey, F. P. (1926). Truth and probability. In F. P. Ramsey (Ed.), *The foundations of mathematics and other logic essays* (pp. 156–198). Routledge, 1931.

Rocca, E., & Anjum, R. L. (2019). Why causal evidencing of risk fails: An example from oil contamination. *Ethics, Policy & Environment, 22,* 197–213.

Rudner, R. (1953). The scientist qua scientist makes value judgments. *Philosophy of Science, 20,* 1–6.

Siemiatycki, J. (2002). Commentary: Epidemiology on the side of the angels. *International Journal of Epidemiology, 31,* 1027–1029.

Suárez, M. (2021). *Philosophy of probability and statistical modelling.* Elements in the philosophy of science. Cambridge University Press.

Part III
What Then When Experts Disagree? Applying Philosophy to Scientific Controversy

Chapter 10
Philosophical Analysis of Some Cases of Disagreement

Detecting Philosophical Bias in Scientific Controversy

We saw in part II that different philosophical biases can motivate different and sometimes conflicting scientific practices and norms. In this final part of the book, 'What
then when experts disagree? Applying philosophy to scientific controversies', the
aim is to put the theory to work in practice and see how a philosophical analysis of
controversy could be done. Most scientific controversies will involve more than one
basic implicit assumption, since philosophical biases of ontological, epistemological, and ethical types are intertwined. Our main purpose here is mainly to show the
principle, so we concentrate on one or two biases for each case. We provide ideas
for further analysis of each case in the sample essay questions. We have chosen a
selection of real-life cases of expert disagreement and philosophical biases, and each
case is based on one or more published articles.

Case Analysis 1

Pollution and Environmental Illness in the Hyde Park Area. Bias about Scientific Procedures and Knowledge as Objective or Relative

Dating back to the 1990s, there was a scientific controversy over whether environmental contamination could be causally connected to the high incidence of illness
in the Hyde Park area. Located in Augusta, Georgia, the Hyde Park area consists of
five low-income African American neighbourhoods surrounded by a heavily industrialised area. In 1993, the Environmental Protection Agency (EPA) carried out a
scientific study of the groundwater, soil, and air in the area to determine the level

of contamination. They found significant levels of heavy metals in the whole area and high levels of toxic substances such as arsenic, chromium, and dioxin in the surface groundwater in two of five neighbourhoods, indicating that the area was heavily polluted. In all these neighbourhoods, there was a high incidence of chronic illnesses such as asthma, skin conditions, rare cancers, and birth defects. EPA had therefore been asked to evaluate the hypothesis of a causal connection between the pollution and the high incidence of illness in the community. After carrying out a scientific assessment, EPA and the Agency for Toxic Substances and Disease Registry (ATSDR) concluded that the chemicals did not constitute a significant threat to the residents' health. This conclusion came in direct opposition to the stand held by residents in the Hyde Park area who, based on their daily lived experience in the area, argued that a causal connection between pollution and illness was not only possible, but also very likely. On this basis, the residents started and won a class action lawsuit where they demanded cleaning up of the area and relocation of the sick inhabitants.

Let us now look in more detail at the arguments and motivations held on each side, by the EPA scientific team and by the Hyde Park's inhabitants. The scientific approach adopted by EPA assumes a classical view on risk as dependent on both hazard and exposure (see Chapter 9). EPA used their pre-defined experimental procedure, which is applied to specific polluted sites. This procedure consists of a classic four-step protocol for scientific risk assessment. The first step of the protocol is hazard identification. Recall from Chapter 9 that the concept of hazard is linked the *potential* to cause an undesired outcome. This is the step to identify whether a certain substance can, at least potentially, cause a certain disease. For instance, if the substance in question is dioxin, the step would be to test whether dioxin can at all be toxic at a certain dose in living animals. The second step is a dose–response analysis. If the substance has health hazards, then how big must the doses be to trigger a toxic effect? This is typically investigated by applying increasing doses of the substance to cell culture and lab animals and recording certain indicators of health in the receiving organism. Once the hazard and the dose–response are established, the third step is to assess the exposure to the hazardous substance. What is investigated in this phase is how much of the toxic substance is found in the environment, and how much of it can reach the population. The physical and chemical properties of the substance are important in this phase, as well as the geology of the polluted area. Does it dissolve easily in water, for instance, or evaporate into air? Could it somehow enter the food chain?

We saw in Chapter 9 that in standard scientific risk assessment procedures, risk is a combination of hazard together with the likelihood of getting exposed to the hazardous substance. This is quite a crucial point, since a hazardous substance that stays inert in deep ocean is not considered risky for human health, no matter the extent and nature of the hazard. In the case of Hyde Park, the scientific team collected 93 soil samples and 14 groundwater samples. They then isolated the chemicals contained in the samples and measured them. Most of the levels that were found fell below toxicity threshold, except for one level of arsenic which was far above the toxicity thresholds, but the level of exposure to the population was considered insufficient to constitute risk. Finally, the risk is calculated in the fourth step by considering both the hazard and the exposure of the population to each substance. The EPA team did not find significant exposure levels, and therefore the final calculated risk was low even though the substances in question are known to pose health hazard for humans.

We see that the standard scientific procedure is based on causal separation, analysis, re-compositions, and experimentation in lab models. There are various relevant aspects that we could emphasise here. First, the scientific assessment was carried out using standardised protocols. Second, the assessment was performed by independent experts who are not stakeholders in the context of inquiry. These are two necessary measures to guarantee the objectivity of the procedures, and such objectivity is a prerequisite for the process to be scientific. This reveals the underlying philosophical bias of objective, neutral, and bias-free observation and we will come back to this in a minute. Before we perform our philosophical analysis, we will have a look at the arguments used by the local population in Hyde Park.

Anthropologist Melissa Checker has studied the residents' perspective on environmental risk by living in Hyde Park for several months. According to Checker's research, Hyde Park residents were highly sceptical of the scientific results generated from EPA's standard protocols for risk assessment. They perceived environmental risk through the lenses of their race and class experience, and they challenged the objectivity of the scientific investigation. In '"But I know it's true": Environmental risk assessment, justice, and anthropology', Checker explains that Hyde Park residents' experience of race and class exclusion was a necessary premise for their stance about environmental illness. Let us look a bit deeper into this perspective.

Hyde Park residents were used to having to fight to get government assistance for basic needs such as water and sewers. Because of this, they had a general mistrust in governmental agencies. In addition, residents experienced that EPA officials were biased against them, were hostile, told them that their health complaints were imaginary, and believed that the real causes of bad health in Hyde Park were rather poor eating, smoking, and exercise habits. On the contrary, African American health officials and politicians were perceived as more supportive of the residents' cause. As a result of their lived experience, the locals interpreted the scientific evaluation made by EPA as biased, and trusted their own daily experience of being poisoned more than the scientifically generated results. Reverend Charles Utley, Hyde Park resident and environmental justice activist, expressed this point by declaring: 'If I set up the test and the test instrument, I can pretty well dictate the outcome'.

According to Hyde Park residents and environmental justice activists, the results from the standardised scientific protocols were untrustworthy for several reasons. First, they objected that EPA tested the effects of one toxin at the time, while Hyde Park was surrounded by six different factories, and located next to the highway and in-between two railroad tracks. Second, they objected to the extrapolation of results from studies using healthy white male workers as a standard (see the discussion about the 'reference man' and WEIRD populations in science in Chapter 4). Third, they criticised that EPA made stereotypical assumptions about the age, size, type of clothing, and sensitivity of the average exposed individual when evaluating the likelihood of exposure to the chemicals. Finally, they contested the measurements of short-term exposure. In sum, the objections pointed to the fact that the normal conditions of standardised protocols were very different from Hyde Park conditions. In one of Checker's interviews, a resident says that when collecting surface soil samples, the testers had actually sampled new dirt that he had imported and put on top of their old, contaminated dirt to protect themselves and their families. The argument is that the local knowledge and perspectives should have been integrated in the standardised protocols to make the scientific assessment better and the results more trustworthy.

The opposing positions of EPA and the Hyde Park residents seem motivated by opposing philosophical biases, related to what counts as best knowledge. The controversy is partly about whether scientific knowledge can ever be purely objective, neutral, and independent, or whether knowledge somehow depends on the knower's standpoint, value, interest, or power (Chapters 1, 4, and 6). On the one hand, EPA assumes that scientific methods and protocols guarantee objective truth and knowledge. It focuses on quantifiable factors such as death rate, level of contamination, exposure, and cost–benefit analysis. The threshold for toxicity is treated as an intrinsic and essential property that remains the same across contexts of exposure, estimated for the standard of some assumed normal and healthy adult man. Chemical interactions are usually not tested, since the default assumption is the principle of additive composition of toxicity, where each chemical is tested in separation and then added in the synthesis. Standard risk assessment, at least as applied by EPA at the time of the Hyde Park case, involves philosophical biases linked to positivism (Chapters 1 and 6), substance ontology, and a view of complexity as mereological composition of its parts (Chapter 7). From this scientific perspective, one can acknowledge uncertainty about the results, but the way to deal with it is by improving the scientific process, or method. One such improvement that was introduced later is the inclusion of more sophisticated models to evaluate more than one environmental stressor at the time (cumulative risk).

On the other hand, Hyde Park residents and environmental justice activists embrace the philosophical bias that all types of knowledge, including scientific knowledge, is somehow relative to perspective and standpoint (Chapter 1), or might even be constructed (Chapter 6) by the experimental design and choice of hypotheses. For instance, they thought that the stereotypical views shared by the scientists about race, class, lifestyle, and health influenced their causal hypotheses and interpretations, searching for other causal explanations than environmental pollution. Local knowledge and information about the unique context were disregarded in the scientific assessment, and EPA and ATSDR used standards that were not representative for the local population, such as single chemical exposures calculated for some version of the 'reference man' (Chapter 4). Hyde Park residents assume a view that knowledge is situated, and that public values and local perception should be given equal weight to scientific facts. A way to face uncertainty in the framework of this philosophical bias is through the precautionary principle. This principle states that precautionary measures should be taken when activity poses a threat to humans, even if causal relationships are scientifically uncertain. The precautionary principle works best within a dispositional and propensity framework, which deals with theories of causal mechanisms and potentials, as opposed to frameworks that emphasises empirical evidence and correlation data (Chapters 8 and 9).

We see, then, that this controversy is motivated by different non-empirical assumptions about knowledge, complexity, causality, and risk. This is why the empirical evidence itself was not enough to settle the disagreement, but instead got dismissed by the opposing position as irrelevant, flawed, or biased. By making the philosophical assumptions more explicit, we can understand why each side would arrive at their conclusions. It also seems that resolving the controversy would require a deeper discussion about the criteria for what counts as objective scientific knowledge and the best scientific method.

Further Readings for This Case

The case is discussed in detail by Melissa Checker in her 2007 article '"But I know it's true": Environmental risk assessment, justice, and anthropology'. In 'Why causal evidencing of risk fails. An example from oil contamination', we discuss how philosophical biases about causality and risk influences the scientific risk assessment of this case (Rocca & Anjum, 2019). The Hyde Park case is presented in a film by Michelle Hansen and Melissa Checker (2016), 'Hyde Park – A Community's Unrelenting Pursuit for Environmental Justice', and in a short documentary by the BREDDL Blue Ridge Environmental Defense League (2014) 'Waiting in a Cesspool'.

Sample Essay Question

Read Checker's paper. Present and discuss the Hyde Park case in light of Chapter 4 about science and power.

Case Analysis 2

Fitness and Adaptation to Climate Change. Bias about Empiricism, Propensities, and Dispositions

Fitness is a key concept in classic evolutionary biology. Indeed, natural selection operates when individuals differ in their fitness. Organisms that are fitter will be selected and therefore survive and reproduce. In 'A causal dispositional account of fitness', philosophers of biology Vanessa Triviño and Laura Nuño de la Rosa have analysed the ontological assumptions that shape the notion of fitness in the fields of sustainable evolution and evolutionary biology. In this case you will learn a bit more about the dispositionalist framework presented in Chapter 8, which is also related to the propensity theory of probability (see Chapter 9). Triviño and Nuño de la Rosa find that fitness can be understood in at least two ways, depending on the philosophical bias involved. Understood dispositionally, fitness is seen as an intrinsic property, or a disposition, of an organism. From a more empiricist perspective, in contrast, fitness is an observable feature or the organism, and can be equated to the number of its offspring. The latter view has been predominant in classic evolutionary biology. We here summarise the analysis and discussion about the concept of fitness presented in their paper 'A causal dispositional account of fitness'. A complete case analysis can also be reviewed in Triviño's lecture 'Philosophical bias and adaptation to climate change. How fit is it, really?'.

Both philosophers of biology and evolutionary biologists are involved in the debate about how to characterise fitness. The discussion has been polarized between those who focus on how fitness should be measured (for instance with genetic or probabilistic approaches), and those who pay more attention to the question of how fitness should be understood as a concept. Clarifying the type of concept that fitness captures, these scholars argue, helps to understand the reason why an organism is able to survive and reproduce in a particular environment.

We said that in classical evolutionary biology, fitness is equated to the actual number of offspring of an organism. Fitness is then characterized in quantitative terms: the higher the number of offspring, the fitter the organism. This is known as the *actualist account* of fitness. One problem for this account is the so-called mismatch problem. The mismatch problem illustrates that in some cases, organisms with more offspring are not the fittest, but the luckiest. A natural catastrophe, for instance, can kill the competitors and predators of the least fit organisms. This way, the least fit organisms would out-survive and out-reproduce the others. In this kind of cases, the authors argue, there is no natural selection involved, but instead pure chance.

The competing view sees the number of offspring as the effect of fitness, rather than as fitness itself. This kind of accounts, also called *dispositional accounts* of fitness, see fitness as a property that can manifest itself as a high number of offspring, but it might not necessarily do so. For instance, female armadillos can arrest reproduction when the environmental conditions are no longer compatible with having offspring. This might happen because of many predators or few resources. According to the dispositional account of fitness, the organism is able to arrest reproduction precisely because it is fit. Indeed, the organism has the disposition to adapt to environmental changes in order to increase its chances of survival. The manifestation of the disposition fitness is in this case not a high number of offspring, quite the contrary. The dispositionalist account distinguishes the property of fitness from its effect, unlike the actualist account that equates fitness with the effect of having many offspring.

There are two different ways of conceiving fitness as a dispositional property, according to the authors. The *propensity interpretation* of fitness equates fitness to a trait or set of traits that characterise an organism. Fitness is the result of having a long peak for instance, or a thick fur. In this propensity interpretation, each trait has a certain value for the purpose of surviving in a certain environment, which can be called the 'fitness value' of a trait. The total fitness of an organism is the result of the fitness value of each trait. An alternative is the *causal dispositionalist interpretation* of fitness, in which fitness is not reduced to the different traits of an organism but is instead a property of the whole organism. According to this view, it's the whole organism that has the capacity, or disposition, to survive, and not one or more traits.

Triviño and Nuño de la Rosa argue that, among the three concepts (the actualist account, the propensity interpretation, and the causal dispositional account) only the last one implies an active role of the organism in the adaptation to the environment. In the *actualist* view, the organism is not seen as playing an active role to adapt to the environment. Instead, the organism is just the carrier of the genes that get naturally selected. In the *propensity* interpretation, according to which fitness is reduced to specific traits, the organism is also seen as the carrier of some traits that happen to become advantageous in some new environmental conditions. In contrast, the *causal dispositionalist* view of fitness allows the organism to actively adapat to the environment and population. Accordingly, fit organisms are organisms that have the capacity to behave as agents, and act in ways that are oriented toward survival and reproduction.

Notice that this third view of fitness as a property of the whole organism is in line with a new concept: *phenotypic plasticity*. This refers to the changes of traits, or phenotype, that an organism undergoes when reacting to environmental change. Because of phenotypic plasticity, it is possible for two organisms of the same species sharing the same genes to have different traits when they are exposed to different environmental factors. Phenotypic plasticity can vary among species because it's grounded on some genes' capacity to produce different traits depending on the environmental inputs. A classic example of plasticity was shown in an experiment in which rats that were licked by mothers at early stages became less aggressive later in life than rats who were not licked. The aggressive trait was manifested in rats that were not licked, regardless of their initial genetics. This means that the trait was expressed in response to an environmental input, and not determined by a certain genetic setup. The phenomenon of environmental plasticity has gained more attention among biologists and there are many examples of it in nature. For instance, the water flea *Daphnia* can manifest a helmet in an environment full of predators, whereas it does not manifest the helmet in more safe environments. The dragon lizard can produce female offspring at high environmental temperature while it produces male offspring at low temperature. It thus seems an advantage if the concept of fitness can accommodate the phenomenon of phenotypic plasticity.

Disciplinary traditions are linked to specific philosophical bias, and we have seen several examples in this book that the same concept is used with different meanings in different disciplines. The case of fitness is no exception. Classical evolutionary biology has traditionally assigned a more passive role of the selected organism, in line with the original concept of natural selection. Traditionally, indeed, there has been no need to see the organism as anything more than the carrier of genetic traits that become advantageous as a consequence of environmental change. For this, the empiricist and actualist view of fitness is most relevant, since it is focuses on the number of offspring. The propensity interpretation, with its focus on the fitness value of single traits, can also match classical evolutionary biology.

The situation is different for new disciplines that focus on the study of phenotypic plasticity. For instance, the discipline of sustainable evolution studies the capacity of organisms to adapt to ecosystem changes produced by climate change. It looks at the capacity of organisms to evolve in a sustainable way, that is, in a way that allows the ecosystem to maintain its biodiversity despite of environmental changes. This question is addressed by studying the phenotypic plasticity of an organism. A causal dispositionalist notion of fitness, understood as a property of the whole organism, is according to Triviño and Nuño de la Rosa better suited for explaining the phenomenon that sustainable evolution studies. This notion of fitness is able to explain how organisms can actively change their traits and adapt to particular environmental changes in order to survive and reproduce. The authors conclude their work by pointing out that the causal dispositionalist concept of fitness, pushed by the study of sustainable evolution, might be adopted to improve discussions in developmental biology as well.

We have here seen that empiricist and dispositionalist biases motivate different notions of fitness. The philosophers of biology who analysed this case argue that which of these notions one assumes will have consequences for which biological theory one can accept. For instance, a change in philosophical bias could support new scientific insights about how organisms develop and survive in response to their environment. Through an explicit discussion about the different philosophical biases motivating the different concepts of fitness, the expert disagreement becomes more transparent.

Further Readings for this Case

An in-depth philosophical analysis of this case can be found in 'A causal dispositional account of fitness', written by Vanessa Triviño and Laura Nuño de la Rosa (2016). Triviño (2020) also discusses the case in the video lecture 'Philosophical bias and adaptation to climate change. How fit is it, really?'.

Sample Essay Question

Watch the lecture by Triviño (2020). In light of this case, discuss how the way we understand the world (ontological bias) influences how we can get knowledge about it (epistemological bias).

Case Analysis 3

Risk Assessment Protocols of Stacked Genetically Modified Plants. Bias about Complexity as Composed or Emergent

Risk assessment of new biotechnologies in agriculture often reveals disagreements within the scientific community. We here consider a disagreement about regulation of the so-called *stacked genetically modified (GM) plants.* Stacked GM plants contain two or more genetic modifications, obtained by breeding two genetically modified parental plants. The disagreement is over whether the evidence that can be generated from traditional risk assessment is adequate and sufficient to predict or even guarantee safe use of stacked GM plants. The dispute is not accommodated by the production of more evidence, which indicates that the disagreement between the two camps is not over the facts, but something else.

Before we proceed to the description of the case, we need to introduce and explain some terminology. Traditionally, farmers have been changing the genetic makeup of crops by breeding them and selecting the offspring for the desired trait(s). This process is called *conventional plant breeding* and its end products, *conventional hybrids*, do not need to be risk assessed before commercialization. Another way of creating plants with desired traits is *transgenic plant transformation.* This is performed by introducing in the genome of a plant some DNA material from a different species (plant, bacterium, or other) with the purpose of introducing a new trait. The end result of transgenic plant transformation is called a *single genetically modified (GM) plant.* For instance, the Bt maize is a single GM maize plant. This type of maize is resistant to certain pests because it produces a toxin, called Bt toxin, which is poisonous for the pests. The Bt toxin is naturally produced by a soil bacterium, Bacillus thuringiensis. The gene encoding for the Bt toxin is transferred to the maize genome through transgenic plant transformation, obtaining a maize plant able to produce a bacterial toxin. There is global agreement that the safety of a single GM plant must be assessed before it is introduced in the market. Such safety assessment includes for instance risk assessment for consumption by humans and animals, environmental risk assessment, as well as the assessment of whether the plant's traits remain stable through generations.

Often it is desirable to have more than one advantageous trait in the same plant. For instance, it might be an advantage that the plant is at the same time resistant to many types of pests, and in addition resistant to the herbicides used to kill weed. One way to do this is to conventionally breed two single GM plants and to select the offspring containing both the transgenic traits. This process of accumulating two or more transgenic fragments in one plant is called *stacking of GM plants*, and the end result of this process is called a *stacked GM plant.*

There is no global agreement on the regulatory demands for stacked GM plants: Should they be considered a product of conventional breeding, and therefore safe? Or should they rather be seen as a new GM entity because they contain a new combination of GM traits? If so, the stacked GM plant should be regulated as a new GM plant, and thereby requires a new safety assessment. European safety assessment agencies think that the latter view is the correct one. European legislation explicitly assumes that 'a stack of two GMOs is simply another distinct GMO, a "new" entity' (European Commission Directorate General for Health and Consumers Evaluation, 2010, p. 60). In order to assess whether the *stacked GM plant* can be granted marketing permission, European Food and Safety Agency (EFSA) (2007) requires data on the stability of the GM plant, potential interactions between the GM inserts, and comparative analyses of nutritional composition and agronomic traits. In contrast, regulatory agencies in countries such as USA, Canada, and Australia adopt the view that since the GM stacks are obtained through traditional breeding techniques, they are no more biologically novel than any other hybrid obtained in the same manner. These agencies therefore require a less rigorous risk assessment scheme before introducing GM stacks on the market.

Scientists have defended either one or the other regulatory scheme. The scientists start from a common ground, which includes three elements. First, they are looking at common safety data, provided partially by independent researchers. Second, scientists agree that single GM plants, if they underwent the necessary safety testing, should be considered safe. Third, they all take conventional breeding to be a safe process. Despite this common ground, the scientists fall into two camps supporting opposite views:

1. **The liberal position**: Stability and potential interactions of transgenic elements in a stacked GM plant can be inferred by information from single GM risk assessment and therefore do not require the generation of new evidence for safety assessment.
2. **The restrictive position**: Some issues cannot be inferred from the risk assessment of single GM parental plants and stacked GM plants therefore require generation of more evidence from a new safety assessment.

In 'How biological background assumptions influence scientific risk evaluation of stacked genetically modified plants: An analysis of research hypotheses and argumentations', Elena Rocca and Fredrik Andersen identify conflicting philosophical biases of an ontological type in the stacked GM controversy. In all the arguments presented by scientists, there were some unstated premises that were necessary to reach the conclusions. Despite their necessity, such premises were always kept implicit. For instance, one typical argument supporting the liberal position 1 is the following:

> *Since the parent GM plants that make up the stacked GM plant have been tested separately, we can take information from these tests and apply it to the stacked GM plant. The only issue that cannot be directly solved from the general safety of single GM plants, is the possible interactions between transgenic proteins in the stacked GM plant. The probability for such interactions is predicable.*

This argument can be spelled out as follows:

Premise 1 (*explicitly stated*): Transgenic protein's behaviour in single GM plants is known.

Premise 2 (*kept implicit*): Genes and their products behave equivalently in single GM plants and in stacked GM plants.

Premise 3 (*explicitly stated*): Transgenic product interaction in stack is predictable.

Conclusion: The behaviour of transgenic proteins in the stacked GM plant can be predicted from the assessment of the single GM plants.

The implicit premise can be considered as the philosophical bias at place here. It's an assumption of 'equivalent behaviour' of entities, meaning that one assumes that the intrinsic properties of an entity (e.g., a protein or a gene) remain the same across different contexts. To explain it with an image, entities are then seen as keys with different shapes. Every key will only fit keyholes with a complementary shape. In this sense, one can say that the fitting of the key is context-sensitive because it depends on the keyhole (the context) that it meets. In the same way, the intrinsic properties of the transgenic insert remain the same in the context of the single GM plant and in the context of the stacked GM plant. We can therefore use the knowledge we have about these properties in one context to predict how the inserted gene will behave in the context of the new plant. We see, then, that any context sensitivity is dictated by the intrinsic properties of the entities, and that such properties are assumed not to change when the context changes.

Philosophically, this assumption is compatible with a mereological composition view of complexity (Chapter 7). Mereological composition is an understanding of complex things as composed of changeless parts, implying that complex wholes can be treated as the sum of the properties of their parts. Atomism is one such view, which we described as a Lego brick ontology. The bricks can combine and compose in countless ways, but their intrinsic properties and essences remain the same throughout this composition. This view of complexity is compatible with a substance ontology, with the assumption that substances (such as proteins, genes, or GM inserts) are more fundamental than processes, and that processes are produced by such substances. It's perhaps not surprising that molecular biologists have traditionally taken this as the default position: that in order to understand processes, such as the interactions between proteins, all we need is to understand the substances themselves in detail. If molecules, proteins, and genes have their properties essentially and intrinsically, then these should remain unaffected by their arrangement or rearrangement in different combinations and contexts.

A similar analysis of the scientific arguments supporting the restrictive position 2 showed that these arguments rely on a different implicit assumption: that entities (such as proteins and genes) might behave differently in parental single GM plants and the new GM stacked plants. In other words, these scientists assume that what entities can do is not entirely determined by their intrinsic properties, but that one must also consider contextual interactions. Context and interactions are integral parts of the entities' properties, and not only external to them. Therefore, one cannot predict properties and behaviour of entities from one context directly to another.

Philosophically, this assumption is compatible with the view of emergence, demergence, and holism. From this perspective, single entities can lose their individual properties and causal powers when combined in a new whole. When a new whole is formed, change could happen in the single parts and not only at the level of the whole. The parts themselves might change as a result of their interaction. A beaver and the environment where the beaver lives, for instance, are both shaped by mutual interaction. Recall from Chapter 7 that this form of emergence is in line with a process ontology, in which entities are understood as being the result of temporarily stable processes, and the best way to understand the behavior of an entity is to study the relations it has with other entities, rather than from studying its internal structure.

With this analysis, we see how conflicting views about the safety demands on stacked GM plants can be partly explained by opposing philosophical biases about complexity. If a scientist tacitly accepts the bias of complexity as emergent, and processes are seen as more foundational than entities, then no amount of empirical evidence on the safety of parental GM plants will suffice to conclude that the stacked GM plant is also safe. And therein lies the disagreement with scientists who tacitly accept a more atomist or substance view of complexity, according to which wholes are thought of as composed of parts that maintain their essences and properties across contexts.

Further Readings for this Case

For a more detailed presentation and discussion of this case, see Elena Rocca and Fredrik Andersen's (2017) paper 'How biological background assumptions influence scientific risk evaluation of stacked genetically modified plants: An analysis of research hypotheses and argumentations'.

Sample Essay Question

In light of Douglas' inductive risk argument presented in Chapter 9, discuss philosophical biases about ethics and value in this case. Which inductive risks do you think the scientists on each side of the debate should consider?

Case Analysis 4

Social Interventions to Improve Child Nutrition in Bangladesh. Bias about Causality as Statistical Difference-Maker or Local Dispositions

The Bangladesh Integrated Nutrition Project, financed jointly by the government of Bangladesh and the World Bank, operated from 1995 to 2002 with the long-term goal to reduce the high level of malnutrition in the country through a series of activities focused on nutrition. The project targeted pregnant and breast-feeding women as well as mothers of children under two years of age, and included growth monitoring, provision of resources for supplementary feeding, and nutrition education. The Bangladesh project design drew heavily on the Tamil Nadu Integrated Nutrition Project in India, which was based on nutritional counselling aimed at making behavioural changes. The design of the Tamil project included an educational program on nutrition targeted at pregnant women, plus a daily budget distributed to the families to cover the expenses for the child's nutrition. The choice to model the Bangladesh project on the Indian one was encouraged by the increasing pressure to use the best available evidence as a guide for rational project-design.

The success of the Tamil project had been documented with randomised controlled trials, in which Indian families were randomly assigned to either an experimental group, which received the nutritional counselling and budget, or a control group, which received no intervention. The children in the experimental group scored significantly higher for health and nutrition related metrics than the children in the control group. Because of the cultural, geographical, and societal similarities between India and Bangladesh, the use of the Indian results to predict the effect of the intervention in Bangladesh was considered scientifically and rationally justified. The use of randomised controlled trials to test and predict the effect of interventions had become the benchmark for good evidence-based policy in the 1990s and 2000s. This trend was dictated by the establishment of the evidence-based medicine paradigm, which postulated a hierarchy of causal evidence for the evaluation of effect of medical interventions. Evidence-based medicine placed scientific methods such as systematic reviews and randomised controlled trials at the top of the pyramid, above observational studies, uncontrolled cases, and any other type of evidence.

The evidence-based paradigm spread beyond medicine and medical interventions to social interventions when teams of scholars (in particular economists from renowned universities) argued that prescriptions for poverty should be as scientifically based as prescriptions for disease, and thus follow the same hierarchy of evidence. In 2004 the leading medical journal *The Lancet*, published an editorial titled 'The World Bank is finally embracing science'. Quoting Esther Duflo, economist and co-founder of the MIT Poverty Action Lab (J-PAL), they write: 'Creating a culture in which rigorous randomised evaluations are promoted, encouraged, and financed has the potential to revolutionise social policy during the twenty-first century, just as randomised trials revolutionised medicine during the 20th'. This approach was seen as a success, and in 2019, Duflo and two of her J-PAL colleagues were awarded the Nobel Prize in economics 'for their experimental approach to alleviating global poverty'. In a press release, J-PAL describes their mission as 'reducing poverty by ensuring that policy is informed by scientific evidence', conducting 'randomized evaluations of innovative policy ideas and programs to identify what works, what doesn't, and why in the fight against poverty; and works with partners to bring the most effective programs to scale' (J-PAL, 15 October 2019).

According to the editorial (*The* Lancet, 2004, p. 731), research carried out by J-PAL had exposed that only 2% of the projects funded by the World Bank over the last few years had been critically evaluated with randomised controlled trials, and that 'billions of dollars are thrown at aid projects without evidence of efficacy'. This is appalling, the editors commented, since such evidence and evaluations 'are public goods, and public accountability surely demands them'. The same editorial warned that 'researchers and policy makers will always have to grapple with the generalizability and replicability of the findings: what works in one country in Africa may not work elsewhere' (Ibid. p. 731). However, this task was presented as fully manageable with the use of hard data and rigorous testing.

From this perspective, it was natural for the Bangladesh project to be modelled on the evidence from the Indian study. Unfortunately, the Bangladesh project failed to produce the desired outcomes. The approach that had given positive results in India ended up failing in Bangladesh. Some experts criticised the evidence-based approach to project-design, centred around randomised controlled trials. They also criticised the initial assumption that the evidence of efficacy in the Indian context could be used as evidence of efficacy in Bangladesh. In 'Will this policy work for you? Predicting effectiveness better', Nancy Cartwright argues that using evidence of efficacy from good studies and pilots to predict whether a policy will be effective is far from a straight-forward task and is sometimes impossible. The reason is that technology and policy typically function by some very local and contextual causal laws and mechanisms, generated by what she calls the 'socioeconomic machine' of that system. The relevant aspects of this socioeconomic machine need to be investigated in detail, and it's difficult to identify in advance whether some key causal aspects differ between contexts.

For instance, it became clear only after the failure of the Bangladesh project that the socioeconomic machine in India consisted in women being responsible for making decisions about food and groceries in the household. In Bangladesh, in contrast, women acted under the supervision of their husbands and mothers-in-law and therefore had less influence on these decisions, something that hindered the success of the project. Moreover, the Indian socioeconomic machine provided grounds for good compliance with the study design. Study participants used fundings from the program to provide complementary nutrition for the new-born child. In contrast, Bangladesh's participants used fundings from the program to provide basic nutrition to all the children in the household. The trouble, explains Cartwright, is that the socioeconomic machine cannot be directly seen and measured, which is what proponents of the evidence-based approach require. Instead, one can only see some elements and build a theory or explanation of how things work.

Let's now analyse these two opposite stances in light of different possible philosophical bias about causality. The two argumentations can be summarised as follows:

1. The Bangladesh project aims to introduce a social measure as a remedy to child under-nutrition. When predicting whether a social measure will have the desired effect, the evidence we should strive to pursue is evidence from randomised controlled trials. Once we have randomised control trials showing efficacy in one context, then we should investigate the applicability to our context through further data. For instance, on can map the similarity between two contexts by measuring factors and parameters and see how they overlap. Theoretical explanations of why and how an intervention worked are to be taken with scepticism since theories can be wrong, and one should avoid interpreting data based on mistaken theories.
2. When predicting whether a social measure will have the desired effect, we first need to understand the social mechanisms at place in the context of interest, as well as how the measure worked in specific local contexts. Such knowledge consists of data and theory, of which randomised controlled trials are only a part. The Bangladesh project should not have been financed without qualitative and ethnographic studies of the societal structure in Bangladesh, and without some theories about how the project got a successful outcome in India.

The reader might recognise that position 1 is based on an empiricist argument. Empiricism is itself a philosophical bias: the idea that we can only know what can be experienced through our senses, and science should therefore only deal with what can be established through observation and measurement (Chapters 1 and 6). Thinking more specifically about causality, we can say that the epistemic priority of randomised controlled trials is motivated by counterfactual and difference-making views of causality (Chapter 8). If causality is understood as something that makes a difference to the effect, then the best way to test causality must be by using studies with comparison and controls. We saw above how the evidence-based methodology for medicine was promoted as a model that should be expanded to social policy. In medicine, the evidence-based pyramid is in fact a hierarchy for evidence of difference-making, where the epistemic value of a method depends on the possibility of comparing two equivalent groups. This possibility decreases from randomised controlled trials, to controlled observational evidence, to case studies and uncontrolled observations. In this position there is also a component of regularity view of causality, since difference-making needs to be evidenced at population level, and not in the single case. One might also use a probability-raising concept of causality. All of these fit well with empiricism, according to which data trumps theory of mechanism. Motivated by these philosophical biases, hard data and large numbers were considered necessary and sufficient for guiding the design of the Bangladesh project.

Arguments in support of position 2 emphasises instead that policy interventions work by local socio-political mechanisms. Understanding such mechanisms is necessary for making predictions, and randomised controlled trials alone cannot provide such understanding. Theoretical understanding of why an intervention makes a difference in a certain context is then necessary for making good use of the evidence from randomised controlled trials, and should be the first type of evidence one seeks. Such arguments assume that difference-making is a symptom of a cause but not the cause itself. A cause is instead thought of as a system's machinery, or potential, that can manifest itself or remain hidden. This intuition fits with a dispositional view of causality. From this perspective, causal knowledge requires mechanistic understanding of the dispositional properties of a system, and its causal powers. Causality is then a local and contextual matter and can involve a unique combination of causal influences. More qualitative approaches are therefore needed to identify and understand the causal mechanisms at place in the single case. In the Indian experience, the cause of success of educating mothers for the purpose of improving child nutrition was the social structure at place, with an emphasis on the decisional power of mothers. Mothers had a potential to improve their child's nutrition because they could make decisions about their meals, but such potential had remained unfulfilled because they lacked the right education and resources. The Tamil project had released such potential. In India there was a disposition at place for mother's education to cause an improvement of child nutrition. Measuring the health outcome in children whose mothers had received nutritional training and funds, as done with the randomised controlled trial, just showed a symptom of such disposition without explaining it. Motivated by these philosophical biases, good policy decision in this case required that one first understands the relevant dispositions in India and then investigate whether the same dispositions were present also in Bangladesh.

Further Readings for this Case

For a more detailed description of the case and further philosophical discussions on causality and methods in science, see Nancy Cartwright's (2012) 'Will this policy work for you? Predicting effectiveness better: How philosophy helps'. For a defence of using RCTs to allocate funding for economic development, see the Editorial in *The* Lancet (2004) 'The World Bank is finally embracing science'.

Sample Essay Question

In light of the case presented here, discuss two contrasting views on general and local scientific knowledge as introduced in Chapter 1. Which of these do you think should be the aim of science, and do you see one as more fundamental?

Case Analysis 5

Vaccine Safety in Health Emergencies and Big Data Pharmacovigilance. Bias about Data as Raw or Theory-Dependent

When a vaccine enters the market, its safety is only partially known. Pre-marketing clinical trials cannot guarantee the detection of all undesired effects, especially on vulnerable groups since they are usually excluded from such trials. Because of this, a system of post-marketing monitoring is at place to identify evidence of suspected adverse effects as early as possible. The science of detecting, assessing, understanding, and preventing adverse effects of medicines (including vaccines) is called *pharmacovigilance*. In this case, we discuss vaccine safety monitoring as an issue linked to big data and philosophical bias about whether knowledge comes primarily from data or theory.

Pharmacovigilance largely depends on electronic databases which collect reports of suspected adverse effects of medicines, and data are often shared between countries in common databases. These types of databases include a huge amount of information, such as VigiBase, the World Health Organization (WHO) global database of individual case safety reports, containing over 20 million reports from patients and clinicians of suspected adverse effects of medicines. These reports are continuously analysed with data-mining approaches with the purpose of identifying medicine—symptom combinations that for some reason are considered of interest for further evaluations. They might, for instance, be reported more often than expected.

A viral pandemic, such as the COVID-19 disease that spread globally in 2020, represents an emergency for pharmacovigilance. In this case, massive vaccination programs were established all over the world that needed close safety monitoring also because of the novelty of the vaccines and scarcity of pre-marketing safety information. This situation generated an extraordinary number of adverse effect reports. Only in the first half of 2021, over 1.100.000 adverse effect reports of COVID-19 vaccines had been shared into VigiBase, something that hadn't happened since the first establishment of pharmacovigilance databases. Processing this amount of data is a challenge because it requires many resources. Even before entering the database, each adverse effect report needs to be digitally transcribed, coded, and structured in a form that can be processed with traditional analytic tools.

We will here discuss two proposed strategies to cope with this vast amount of data. The first strategy focuses on improving technology for data mining, while the second focuses on interdisciplinary teams and a closer collaboration between pharmacological experts and data scientists. These two strategies aren't necessarily mutually exclusive, yet one can see that manufacturers and regulatory agencies have different preferences. Manufacturers tend to hold an overly optimistic attitude toward digitalisation, automatisation, and the development of more sophisticated data-mining algorithms and AI technology: 'Using smart technology to manage the [...] process not only simplifies what can be laborious and time-consuming work for humans, but can also help to reassure members of the public who are concerned about the safety of newly developed drugs' (Pharmafile, 2021). On the other hand, some regulatory agencies as well as independent researchers have warned that technology cannot replace the need for human expertise, but can be helpful if combined with the interdisciplinary work of medical doctors, pharmacologists, and data scientists.

The tension between these different strategies mirrors the philosophical debate on big data, and the different roles of data and theory in science (Chapter 6). The first strategy assumes that big data and automation improves the objectivity of the inquiry, which fits a positivist and strict empiricist philosophical bias (Chapters 1 and 6). The second strategy assumes that data are context- and theory-dependent, which fits the philosophical bias of perspectivism (Chapter 1) or constructivism (Chapter 6). From this point of view, the knowledge and theoretical perspectives of the data curators are situated rather than objective.

The positivist, technology-focused take on big data pharmacovigilance is exemplified by the large amount of resources spent during the COVID-19 pandemics to increase the capacity to process as much data as possible from as many sources as possible. Such efforts included the development of an algorithm that could automatically code the adverse effect reports and translate them into standardised terminology that could make them searchable and retrievable from databases. Moreover, technological development was aimed at automatic translation of reports from different languages into English, and automatic removal of patient-sensitive data so that the reports could be available for analysis without breaking privacy protection laws. More remarkably, perhaps, the horizon of potentially useful data expanded beyond the already big databases of spontaneous reports, and efforts were made to merge different sorts of data, for instance from health registries, claim registries, and even experiences shared in social media. This strategy involves significant investments of resources from the data science expertise, because joining different registries, databases, and health records requires further standardisation and a common language for coding. Yet, the amount of resources needed was not discouraging. On the contrary, the European Medicine Agency recognised that 'Big Data [...] offers major opportunities to improve the evidence upon which we take decisions on medicines' (European Medicines Agency, 2020, p. 3) and set up a Big Data Taskforce to develop the necessary technical skills, capacity, and tools for the joint analysis of different type of data sources.

This effort to increase both the amount of data and the automation of data processing is not surprising given the role and status that big data has gained in the last decades. Back in 2008, the technology magazine *WIRED* published an article titled 'The end of theory: The data deluge makes the scientific method obsolete', making the following optimistic claim: 'With enough data, the numbers speak for themselves, correlation replaces causation, and science can advance even without coherent models or unified theories' (Anderson, 2008). Although this statement is quite provocative, it implies a philosophical bias that is not uncommon in science: that science should deal more with data and less with theory. Cherry Murray, who established the Institute for Applied Computational Science at Harvard in 2010, stated that 'the connectivity of the Internet and the microelectronics revolution are enabling us to collect, store, interact with, and learn from massive streams of raw data' (Perry, 2014). We see that Murray here expresses both the idea that data are raw, meaning that they have not been processed by single individuals, and the idea that such raw data can give us knowledge. These are powerful statements to make about the promises of big data, suggesting that large amounts of raw data have the potential to increase the reliability of scientific, political, and sociological models because they represent the death of theory and subjectivity. Recall that theories lean on general knowledge, understanding of phenomena, judgement, expertise, and perspectives of single epistemic agents or communities, which are always incomplete and therefore seen as unreliable. If we have enough data of sufficient quality, the reasoning goes, we will be able to detect that one vaccine is associated with a certain symptom, even if there is not yet any pharmacological theory or understanding of the dispositions or mechanisms involved; of how the symptom might be brough about by the vaccine.

We can see then that the confidence in big data and AI technology as a superior tool for scientific enquiry is based on the combination of two philosophical biases: the empiricist idea that knowledge comes primarily from what we can experience via our senses (Chapter 1), combined with the idea that subjectivity gives a distorted and partial view on such truth, and that science offers the best tools because it can at least attempt objective and bias-free observation. In Chapter 6, we placed positivism, on one side or the debate, in opposition to relativism and constructivism. The second camp, opposing the big data and AI strategy, argues that no knowledge or data can be independent of context, theory, or perspective. We have explained how Norwood Russell Hanson argued that all data are theory-dependent. We also saw that Sabina Leonelli warns scientists against believing that data are raw, context-free, or neutral (Chapter 6). The same has been argued specifically for the case of big data pharmacovigilance and COVID-19 vaccines. Let us illustrate this with a specific case presented in 'Monitoring the safety of medicines and vaccines in times of pandemic', written by Elena Rocca and pharmacovigilance expert Birgitta Grundmark.

When entering adverse effect reports in a pharmacovigilance database, pharmacovigilance specialists need to code the name of medicines and vaccines with a standardised international classification. One of these classifications follows the WHO Drug dictionary. This dictionary classifies medicines in different groups based on various criteria, such as their pharmacological effect, indication for treatment, or metabolic pathway. These groups are called Standardised Drug Groupings (SDG). Such grouping criteria are relevant for different purposes, and we will now consider the use of the SDG resource for safety monitoring analysis.

> Imagine for instance that I suspect that a medicine X causes a certain adverse effect because it inhibits receptor R. Being able to retrieve a group of safety reports containing medicines similar to the medicine of interest X, in that they all inhibit receptor R, is important. It allows me to check, for instance, whether there is a significant correlation with the adverse effect of interest in the total number of reports at the SDG group level. This gives an indication to support (or not) the hypothesis. (Rocca & Grundmark, 2021, p. 139)

A database user can make as many group queries as there are SDGs available. The more relevant the SDGs are for a certain purpose, the more efficient the data mining and analysis. Let us now look at SDGs in the specific case of analysing the adverse effect reports for COVID-19 vaccines. WHO Drug curators continuously update and create new SDGs. To facilitate the safety monitoring of COVID-19 vaccines, they grouped the different types of vaccines according to different criteria, so that reports on vaccines belonging to a certain group could be retrieved together and analysed simultaneously. But how should adverse effect reports from different types of COVID-19 vaccines be grouped together? This information does not come from the data themselves. Instead, an evaluation is based on the curator's judgement of how clinical and pharmacological interactions are best represented and investigated.

For instance, WHO Drug curators suggested that one way of grouping COVID-19 vaccines is to classify them based on the vaccine platform. This means that if a pharmacovigilance assessor finds that there is an unexpectedly high number of reports naming a type of RNA COVID-19 vaccine together with a certain symptom, they can retrieve all the reports with RNA COVID-19 vaccines available at the time and check whether the symptom is reported unexpectedly frequently also at group level. This is done because curators believed that it makes clinical sense that RNA-based vaccines might interact with the body in different ways than other types of vaccines, which can result in properties and potential effects that are common to RNA-vaccines as a group. This idea, that the type of vaccine platform has something to say about the adverse effects it may provoke is indeed a theory, anchored in clinical and pharmacological reasoning. There are many other ways to classify COVID-19 vaccines, but just some of these ways will be considered clinically and pharmacologically relevant by the curators and therefore be made available as a group query in the SDG tool.

Scholars making this point argue that, far from being the end of theory, data must be made usable, which requires considerable labour and expert judgement. As a result of this, big data turned into knowledge is clearly dependent on theory and the construction of theory-informed categorisations and classifications. This side of the debate emphasises the need for interdisciplinarity between data science and pharmacological expertise, and warns against the one-sided investment of the majority of resources in data science, programming, artificial intelligence, and other purely technological advances. From the opposite bias, of positivism, such investment is the best way to go and will help generate objective and reliable knowledge based entirely on large amounts of data. We see, then, that this debate is motivated by conflicting epistemological biases, about how get knowledge. Specifically, it is a disagreement about the role of data and theory for generating scientific knowledge.

Further Readings for this Case

For a further philosophical discussion on types of evidence and decisions in drug safety, especially during the COVID-19 pandemic, see Elena Rocca and Birgitta Grundmark's (2021) article 'Monitoring the safety of medicines and vaccines in times of pandemic: Practical, conceptual, and ethical challenges in pharmacovigilance'. For a broader discussion and reflection on big data science, see 'The challenges of big data biology' by Sabina Leonelli (2019).

Sample Essay Question

Discuss the ideal of big data pharmacovigilance presented in this case in light of Francis Bacon's inductive method, as described in Chapter 2. Consider the idea that induction and deduction can be understood as two philosophical biases.

Case Analysis 6

Biodiversity Mapping and Forest Conservation. Bias about Values and Scientific Methods

Sustainable forest management is an area that requires interdisciplinary collaboration and expertise. However, because of the diverse economic, environmental, and political interests in the forest, this field is full of controversies and expert disagreement. When conflicting interests are involved, scientists worry about bias. In this case, however, we will see that the biases related to interest can have ontological and epistemological motivations and implications, affecting the methods and tools used to map biodiversity in the forest. Two methodologies for biodiversity mapping in the field of forest management is analysed for philosophical bias, based on the work by two researchers at the Norwegian Institute for Nature Research (NINA). In 'Segmented forest realities: The ontological politics of biodiversity mapping', Håkon Aspøy and Håkon Stokland argue that the two methodologies rely on competing ontologies, and that the different ontologies have different political implications. Here, we give an overview of this case, which can be read in detail in Aspøy and Stokland's article. Our discussion of philosophical biases that could motivate the different methodologies are not restricted to those discussed by Aspøy and Stokland.

There has been a long-lasting discussion about scientific knowledge related to forest conservation in Norway. Specifically, there are controversies about the practice of biodiversity mapping, which has a central role in the production of such knowledge. Biodiversity mapping involves observing the types of species and habitats in fieldwork and plotting it into a map. Mapping is important both for environmental governance in the public sector and for environmental certification of private forest properties.

Aspøy and Stokland analyse a specific debate about the use of two different methods for biodiversity mapping, which they argue are based in different ontologies, or 'forest realities'. The first method is a procedure to map Norwegian coniferous forests, developed by a group of conservation biologists that called themselves 'Siste sjanse', meaning 'Last chance'. The procedure was given the acronym SiS and was aimed at saving biodiversity-rich areas of the forest from being cut. The second procedure, called 'Environmental Inventories in Forests' (EiF) was developed by the Norwegian Ministry of Agriculture and promoted as a science-based method. Because of this, EiF became the dominating methodology for biodiversity mapping in Norwegian forestry. At the same time, the Ministry gradually decreased the use of the SiS methodology, despite protests by some NGOs who argued for the necessity of a variety of mapping tools. Both SiS and EiF are fieldwork-based methodologies and they both focus on identifying 'woodland key habitats' that offer good conditions for biodiversity. Despite these similarities, the methodologies are different in their core and reflect different ontologies, of forest realities, and different epistemologies, on what makes a procedure for mapping biodiversity in forests *scientific*.

> SiS and EiF enact forest realities that are quite similar but differ in some crucial respects. SiS placed great emphasis on finding woodland key habitats, while EiF disputed that woodland key habitats could simply be found. This was related to another aspect of the forest reality enacted by EiF. Unlike SiS, which presumed that biodiversity is often concentrated in hotspots, EiF assumed that biodiversity occurrences are distributed more evenly and throughout larger forest areas. The emphasis in EiF on less variation made the enacted forests more predictable, comprehensible and standardized. The emphasis in SiS on variation and complexity, on the other hand, contributed to the enactment of forests that were difficult to know and predict. (Aspøy & Stokland, 2022, p. 123)

If we look at philosophical biases discussed in this book, this controversy includes a range of contrasting views: data versus theory, objective versus situated knowledge, empiricism versus perspectivism (all presented in Chapter 6), reductionism versus emergence (Chapter 7), causal mechanisms versus observed correlations (Chapter 8), general versus local knowledge (Chapter 1), and standardisation versus context-sensitivity (Chapter 9). For instance, SiS seems to be more in line with epistemologies and ontologies that emphasise theories of causal mechanisms, local and context-sensitive knowledge, intrinsic dispositions, and single propensities. In contrast, EiF emphasises positivist ideals such as standardisation, objectivity, neutrality, and quantification. These ideals are in line with strict empiricism, positivism, and priority of correlation data over theories of causal mechanisms. Let us explain this in a bit more detail.

In SiS, the practitioner would evaluate the presence of key elements that are known to favour biodiversity, such as dead wood or water ponds, and the presence of rare species, the idea being that such species thrive in biodiversity hotspots and could therefore be treated as signal species. The mapper would focus on the ecological interactions with the larger ecosystem, and the result would often be the identification of relatively large key habitats. The idea was that finding the borders of the key habitat was particularly difficult especially because of the uncertainty around what was needed for a certain species to survive. That is why this work required a good deal of expertise, qualitative evaluations, and precaution. In other words, SiS acknowledged the difficulty of finding key habitats and the challenges of knowing and predicting forest dynamics. It therefore emphasised variation and complexity. Philosophically, the SiS strategy fits the dispositionalist ontology, according to which any causal process is a result of complex interactions with a unique combination of properties involved. What something does in one context will therefore not be the same as it does in a different context, because there will be different properties and interactions involved. On this view, any standardised approach would fail to accommodate local variations and uniqueness.

On the other hand, Aspøy and Stokland explains, EiF did not have a focus on biodiversity hotspots and the Ministry was sceptical to the idea of signal species. It postulated instead that key habitats could be found relatively easily by an objective and neutral observer by counting key elements (water ponds, dead trees and so on), which in their view were distributed evenly throughout the forest. This is why EiF emphasised standardisation and predictability over variation, complexity, and precaution. EiF also downplayed the importance of preserving some areas where indicators of biodiversity were found and stressed the need of conserving key elements in a uniform way throughout the forest. The EiF strategy reflects the positivist ideals of knowledge, which promote the recording of empirical facts, quantification and comparisons of these, and observed repetition of same cause, same effect, under some standard, normal, or ideal conditions. From this perspective, observed patterns of regularity offer a solid ground for categorisation and prediction.

A further difference between SiS and EiF concerns the mapper's role for the decision on which forest areas should be conserved. In SiS, the mapper had the role of making recommendation about forest management. The basic idea was that decisions about management were biological or ecological decisions, and therefore they required the evaluations of a biologist who was given the authority to make evaluations and influence the practices of the forest owners. Philosophically, this is in line with the idea that causal knowledge must include theories of mechanisms, and the assumption that empirical data are theory-dependent. This would require scientific expertise and interpretation of the data as an integral part of the mapping process. In contrast, in EiF there is a strict division between descriptive responsibilities, which would belong to the mapper in the field, and normative responsibilities, which would belong to a group of stakeholders including forest owner, biologists, the private sector, and public forestry administrators. In other words, decisions about forest conservation were not based on biological and ecological concerns. The mapper's role was therefore limited to neutrally mapping quantitative information during the fieldwork without giving any advice or management proposal. Philosophically, this is in line with Francis Bacon's empiricist ideal, according to which the role of the scientist is to be a neutral and bias-free observer who adds nothing to the observation in terms of focus, interest, or agenda (Chapter 2). The aim is to record and classify the data through observation alone, without the aid of theory.

Different political communities have supported the two different methodologies, and the result was a coalition of NGOs and the environmental segment supporting SiS, while a coalition of the government and the forestry segment supporting EiF. This is not surprising, since we saw that the methodological struggle has affected the ability of these communities to influence decision-making. Ecology and the environmental segment are empowered in SiS, while the socio-economical segment is empowered in EiF. SiS supporters argued that identifying key areas for conservation required trained biologists, while EiF supporters accused the counterpart of proposing scientifically doubtful procedures in order to maintain substantial influence in the decision-making process.

Note that the different methodologies give different evaluations of the same forest areas. For example, forests in the Norwegian region of Telemark were evaluated as having a high significance for biodiversity when SiS was used. The consequence would be that large areas of the forest should be protected. On the other hand, the same forest was evaluated as less significant for biodiversity when EiF was used. This case is thus an example of how ontological and epistemological biases are intertwined with social and ethical interests and might play a major role in shaping the scientific process and scientific results, together with the empirical evidence.

Further Readings for this Case

Further philosophical discussions on this case can be found in Håkon Aspøy and Håkon Stokland's (2022) article 'The ontological politics of biodiversity mapping'.

Sample Essay Question

Discuss the role of community for what counts as scientific knowledge and method. Specifically, discuss the case in light of Thomas Kuhn's notion of normal science, and contrast this with the concept of post-normal science by Silvio Funtowicz and Jerome Ravetz, as introduced in Chapter 3.

Case Analysis 7

Risk of Foetal Malformations from an Anti-epileptic Medicine. Bias about Inductive Risk, Values, and Probability

Epilim is a medicine used to treat epilepsy that contains an active molecule called *sodium valproate*. Sodium valproate is a widely used medicine that has been on the market since 1974 with different commercial names. It is used for the treatment of certain types of epilepsy, but also for treating bipolar disorder and migraines. The use of sodium valproate during pregnancy has been associated with a series of specific malformations of the foetus, which since the mid-1990s is known with the name of 'valproate syndrome'. These include serious and rare defects in the development of the nervous system. One of the birth defects associated with the medicine, for instance, is spina bifida, in which the spinal cord does not form properly, leaving a part of nerves and spinal cord exposed in the new-born's back. Sodium valproate has been at the centre of a scandal involving legal lawsuits and class actions in which parents of damaged children accused the manufacturers and drug authorities of being aware of the risk of foetal malformations long before these started to be clearly communicated to pregnant women who were treated with the drug.

Today, the leaflets of medications containing sodium valproate indicate that the medicine should not be used during pregnancy because of high risk of malformations of the foetus. However, this warning wasn't always there. Although the first description of the foetal valproate syndrome dates back to 1984, and although concerns about foetal toxicity were present already at the launch, the public recognition of the syndrome started only in the mid-1990s. The decision of communicating the risk of a certain side-effect is a country-specific decision, made by national drug authorities after consultancy with expert groups. Interestingly, different expert groups (and consequently different drug authorities) made different evaluations on whether the evidence available at different times was enough to motivate a clear communication of risk.

Let us look more in detail at the accumulating evidence for a causal association between the medicine and the malformations.

> Launch 1974: Sodium valproate belongs to the class of anti-epileptic medicines, which are medicines used to treat epilepsy. In 1974 there was an indication that the use of some medicines from the same class was correlated with increase of foetal malformations. Indeed, all anti-epileptic at the time were marketed with a generic warning to fertile women, stating that the potential risks should be weighed against the potential benefits.
>
> Soon after the launch: Foetal toxicity in different animal species was demonstrated.
>
> 1982: By 1982, several case reports of spina bifida in babies born from mothers who used the medicine in the first trimester of pregnancy had been collected. The number of case reports suggested an estimated 1–2% incidence of spina bifida in new-borns who were exposed to the medicine during foetal life. This frequency is very high if we consider that in the US the background incidence of the malformation was only 6 in 100.000. However, it was commented that each of these early signals contained bias, confounding, and technical issues that made it difficult to evaluate whether the association was due to the medicine or confounders.
>
> From 1985: Initiation of human studies of different types, with progressive intensification. For instance, case–control studies were performed. In this study design one looks for a common cause among a group of new-borns with the same defect by comparing the medical record of the mothers against a control. Another type of clinical study that was initiated is cohort studies, in which a group of women is observed for a period of time in order to find a correlation between the exposure to a possible toxic substance and the health outcomes in the new-born.

How was this initial evidence evaluated? Let us consider the evaluations by two different expert groups: clinical teratologists and drug agencies. When comparing the scientific publications available in the 1980s with the concurrent teratology textbooks and medicine's data sheets, we find a striking divergence in the way scholars and different drug agencies communicated findings and uncertainties.

Clinical teratologists specialise in counselling patients about exposures that may cause birth defects, be it medicines, infections, physical agents, or chemicals. Clinical teratology is taught in medical education and specialisations. Crucially, textbooks in clinical teratology started to list valproic acid as a teratogen (= cause of foetal abnormalities) for animals and humans as early as 1985. For clinical teratologists, then, the adequate communication of details about the potential undesired effect, such as details about the observed malformations, comments about dose–response, and relevance to humans, could have been helpful for clinical decision-making. In contrast, the British Department of Health and Social Security (DHSS) did not update the datasheet of medicines containing valproic acid, such as Epilim, until 1990. For the first time in 1990, a specific warning regarding the risk for foetal malformations in humans appeared, accompanied by a quantitative estimation of the risk. The warning reads as follows: '…There have been reports of foetal abnormalities including neural tube defects in women receiving valproate during the first trimester. This incidence has been estimated to be in the region of 1%.' The five-year-gap delay in the update of the medicine's data sheet was one of the concerns that sparked the protests and lawsuits by parents of damaged children.

We can now consider which philosophical bias could have been made explicit and communicated by the different expert groups in order to increase the transparency of the evaluations and make this discrepancy more understandable, and perhaps acceptable. For a theoretical introduction to some of the possible philosophical biases about risk, see Chapter 9.

First, there might have been different value judgements between clinical teratologists, whose task is to support the clinical communication with single patients, and medicine agencies, whose task is to give a public warning in the form of an updated leaflet. It is reasonable to think that medicine authorities focus on the effects that regulatory actions will have on a large scale. There might have been an inductive risk argument by the medicine authorities, since a public warning in the medicine leaflet can work as a motivation for many epileptic women to discontinue the therapy during pregnancy, without a medical consultation. Because of this, medicine agencies may be particularly cautious to avoid a false positive, that is, they may be mostly worried about warning against the risk of an undesired effect that might in fact not exist. It is known, for instance, that epileptic seizures themselves can cause serious damage to the foetus. The control of such seizures during pregnancy is a delicate matter, which requires balancing the risk of exposing the developing foetus to anticonvulsants against the risk of seizures due to discontinuation of the therapy in early stages of pregnancy. These considerations can motivate the medicine agency's requirement for the accumulation of more definitive evidence supporting a causal link between valproic acid and foetal malformations.

Second, the two expert groups might have based their risk evaluations on different philosophical bias about probability. Clinical teratologists acknowledged that valproic acid has an intrinsic property to provoke foetal malformations. This intrinsic property had manifested itself only in animals, yet animal evidence combined with the single case reports and the similarity to other teratogenic medicines pointed to an intrinsic potential of the medicine to also hurt humans. This is compatible with a propensity understanding of probability. Saying that there is a chance that valproic acid hurts a human foetus means that the medicine has a dispositional propensity to do that, which is here generated by its physical properties and their mutual manifestation partners (in this case, the physical properties of the person receiving the drug). This propensity might not be observable as a long-run frequency, for instance because only very few individuals will have properties that are mutual manifestation partners for the harmful effect, but the disposition of the drug would still exist and might manifest itself in the single case.

In contrast, a frequentist view on probability does not allow to acknowledge, say, a 1% risk of a human foetus to be hurt by valproic acid before one is reasonably sure that the effect has been observed one in hundred times. This observation cannot be based on single cases or on particularly weak studies in which confounders are not controlled. As an additional problem, a frequency of incidence is technically difficult to observe with a prospective study, given the rarity of the foetal exposure in the first critical semester of pregnancy. If a frequency of outcome is not observed, then probability cannot be assigned within this philosophical framework. A sceptical view toward the evidence available in the 1980s as evidence of a risk worth communication is then compatible with a frequentist theory of probability. Medicine agencies could have offered a better and more transparent argumentation for their evidence evaluation by explaining why a frequentist view on probability is more apt to a regulatory decision than the propensity view adopted in clinical pharmacology textbooks, for instance by using inductive risk arguments.

Further Readings for this Case

To learn more about the case you might find it interesting to read Asher Ornoy's (2009) article 'Valproic acid in pregnancy: How much are we endangering the embryo and fetus?'. The case is also discussed briefly in Elena Rocca, Ralph Edwards, and Samantha Copeland (2019): 'Pharmacovigilance as scientific discovery: An argument for trans-disciplinarity'.

Sample Essay Question

In light of this case, discuss how making philosophical biases explicit could affect risk communication from the scientific community to the public. Consider also what might make risk acceptable or unacceptable, for instance for different stakeholders.

Case Analysis 8

Sustainable Pig Farming and Management of Viral Infections. Bias about Substance and Process Ontology

In the field of life technoscience there is a long-lasting debate about the development and assessment of technologies. On the one hand, there is a narrative of control and precision that can be achieved by zooming technological fixes into the molecular level of the issue. On the other hand, there is a narrative of scepticism toward reductionist solutions which do not acknowledge real world complexity. One example of such debates is the attempt to control the spread of vaccine-resistant viral infections in high-density animal farming by using gene-editing techniques. Sometimes farmed animals die in large numbers because of viral diseases that cannot be prevented by vaccination. In these cases, one of the proposed solutions is to edit the animals' genome, so that the animals cannot be attacked by the virus, for instance because their cells lack the molecule that the virus uses as a gateway to infect the body.

Philosopher and biochemist Stephan Guttinger discusses a case of infection management in high intensity pig farming and analyses it for philosophical bias. The complete case analysis can be reviewed in Guttinger's lecture '"Frankenswine", genome editing, and the question of sustainable pig farming'. In the case described, molecular biologists have removed from the pigs' genome the molecule that serves as target for a virus producing a lethal respiratory syndrome. In other words, scientists have tried to protect pigs from a disease by creating mutated pigs whose cells lack one protein. The deleted protein is necessary for the pathogen, a virus, to anchor to the cell surface and enter it. The disease is called Porcine Reproductive and Respiratory Syndrome (PRRS) and is a pig-specific syndrome with global spread. Globally, PRRS has been causing several billion US dollar losses because of reproductive failure and the high mortality rate due to respiratory issues including pneumonia, as well as increased death of embryos and new-born pigs. Because of this, there is a huge interest in stopping the global spread of the syndrome, but to this date vaccination has not been a reliable and effective measure.

PRRS is caused by a type of virus that contains RNA as its genetic material. RNA viruses are common and in humans they can cause for instance influenza and COVID-19. By binding to specific molecules in the cell surface, RNA viruses enter the cell and reproduce by transcribing the viral RNA and synthetising new viral proteins. In the case of the PRRS virus, the target cells are pigs' macrophages, a type of white blood cell that is part of the immune system which contributes to the first line defence against pathogens, also called 'innate immunity'. The PRRS virus enters the pig's macrophage by binding to CD163, a molecule that is attached to the macrophage's surface.

With modern genome editing techniques it is possible to delete the part of the pigs' genome that is responsible for producing CD163, and according to molecular biologists this can happen in a precise and clean way, at least in principle. The macrophages of the mutant pigs, which are lacking the CD163 gene, would not produce the CD163 and therefore could not be attacked by the virus. As a result of such gene editing, the mutant pigs did not get infected from the virus and did not have important side-effects, at least in experimental evidence. Shouldn't these results increase our faith in biotechnology and encourage an implementation in large scale? Herein lies the expert disagreement, where one side says 'yes, definitely', while the other side is far more sceptical.

According to the scientists behind the gene editing technology, the positive results show its success and potential to solve a problem for pig farmers worldwide. From this perspective, all biological systems including viruses are treated as molecular machines. Biological properties are defined by a setup of molecules, and in particular by the genomic sequences. If the genome sequence is modified, then, the organism and its capacities will also change automatically and by necessity. On this view, a fix at the molecular level works as a clean and time-effective solution.

On the other hand, there is a narrative of scepticism toward solutions of this type. Some virologists refuse to see viruses as genetic entities that can be described as a defined sequence of DNA or, in the case of the PRRS virus, as a defined RNA sequence. The scepticism comes from the fact that the genetic material of viruses very often mutate: Each newly produced virus contains one to three changes in its genome. One could say that each single pig cell that is attacked by the virus produces a cloud of mutants, rather than a series of copies of the original virus. Some contemporary virologists argue that we should stop thinking of viruses as defined molecular entities that can be described by a certain DNA or RNA sequence. Rather, viral particles should be seen as a cloud or spectrum of mutants that interact dynamically. Another important point, they argue, is that the evolutionary success of the mutant clouds is co-determined by its surroundings, which in this case is the internal environment of the pig's cell. Modern virology then challenges the gene-focused approach of molecular biology. Since viruses are dynamic and context dependent entities in continuous change, how can we exclude the possibility that the PRRS virus will start to work in CD163-free pig cells? On the contrary, some virologists warn, one should expect that the virus will start attacking its target through a different molecule. Gene-editing is then seen only as a short-term fix which will not be able to solve PRRS in farmed pigs in the long run, but at best just manage it for a limited period.

A third expert group involved in the discussion, in addition to molecular biologists and virologists, are veterinarians. Some veterinarians point out that viruses adapt to the ecosystems that they live in, and that viral infections cause higher mortality rate when they infect a large and dense population, while the symptoms become milder when population is sparse, and it is more difficult for the virus to find a new host to infect. Experts from the veterinary field have therefore often argued that the real problem of PRRS are not the viruses, but the farming practices, with high intensity and large herds. If the aim is to reduce the mortality of PRRS infection, one should modify the practices of animal husbandry, rather than making an intrinsic change in animal genetics.

We see, then, that although the experimental evidence shows that genome editing has worked in creating animals that are immune to the infection, such evidence is interpreted differently by different experts. This is an indication of a deeper, underlying disagreement, which is not explicitly discussed. Let's consider some philosophical biases that might be the source of disagreement.

Some philosophers of science and technology have argued that the attitude to technoscience held by molecular biologists is often based on an ontological bias of bottom-up causality, in which molecules are seen as the stable unities that originate life. The opposite view, sceptical to solutions exclusively based on biotechnology, assumes instead a top-down, relational, and deeply contextual view of causality (see, for instance, Wickson & Wynne, 2012). Top-down versus bottom-up causality is also linked to philosophical biases about complexity, and to whether one sees the smaller entities as substances that keep their properties or essences intact across contextual change and interactions. We saw in the discussion about substance and process ontology (Chapter 7) that there is a tension between reductionism and emergence, which gives different expectations about how much one needs to consider environmental influences. Mereological composition, and atomism, suggests that the parts behave like Lego bricks that remain unchanged by their context, which is in line with the optimism toward the gene editing technology. In contrast, if the parts themselves might change depending on which other parts they are combined with, then the result is not easy to predict, and one might adopt a more precautionary stand toward such bottom-up technology (see also the discussion of case 3 above).

Philosophical biases about processes, substances, reductionism, and emergence are not the only source of the disagreement. There are other divergent biases involved, including ethical ones. Different socio-economic interests can influence the view on biotechnology, for instance about what makes certain risks acceptable, or which inductive risks are worth considering (see Chapter 9).

In his case analysis, Guttinger makes a point worth noticing. Since science-based measures would have to be applied to the practice of animal farming, one should include the arguments of field practitioners in the philosophical analysis. This means that one should consider the cutting-edge in farming practices when making decisions about how to move forward to solve this problem. Is the field itself progressing toward a bottom-up assumption of causality, or toward a top-down, contextual view of reality? This is a complex question that can be answered to by looking at the dominant arguments in the scientific environment and the trends in the practice. Interestingly, answering this question could require a collaboration between philosophers, scientists, and practitioners, thus setting the ground for genuine trans-disciplinarity. Considering the trends in farming practices, Guttinger notes that the Soil Association has criticised the modern farming standards. The critics have recommended that a solution to pig epidemics requires an adjustment of the farming processes. For instance, one should stop introducing new pigs to the herd for at least 200 days. Some practitioners seem therefore to push for a process-adjustment, requiring the coordination between different players, such as farmers, veterinarians, and health officials, rather than a pure molecular approach using gene editing technology. Guttinger concludes that, once we factor in findings from virology and veterinary, and if we put all of this in context of actual farming practices, a relational worldview seems to be the better-supported view.

Further Readings for this Case

Stephan Guttinger (2020) describes the case in the recorded lecture '"Frankenswine", genome editing, and the question of sustainable pig farming'. Further discussions on the view of molecules as processes can be found in Guttinger's (2021) 'Process and practice: Understanding the nature of molecules'.

Sample Essay Question

Watch the lecture by Guttinger (2020). Present and discuss the case in light of the value hierarchies introduced in Chapter 5: anthropocentrism, zoo-centrism, bio-centrism, and eco-centrism.

References

Anderson, C. (2008). *The end of theory: The data deluge makes the scientific method obsolete.* WIRED. https://www.wired.com/2008/06/pb-theory/. Published on 23 June 2008.

Håkon Aspøy & Håkon Stokland (2022). Segmented forest realities: The ontological politics of biodiversity mapping. *Environmental Science & Policy, 137*, 120–127. https://doi.org/10.1016/j.envsci.2022.08.015

BREDL—Blue Ridge Environmental Defense League. (2014). *Waiting in a Cesspool.* https://bredl.org/resources/waiting-in-a-cesspool/

Cartwright, N. (2012). Will this policy work for you? Predicting effectiveness better: How philosophy helps. *Philosophy of Science, 79*, 973–989.

Directorate General for Health and Consumers Evaluation of the EU. (2010). *Evaluation of the EU Legislative Framework in the Field of Medicated Feed, Final Report.* https://food.ec.europa.eu/system/files/2016-10/gmo_rep-stud_2010_report_eval-gm.pdf

European Food and Safety Agency (EFSA). (2007). Guidance document for the risk assessment of genetically modified plants containing stacked transformation events by the Scientific Panel on Genetically Modified Organisms (GMO). *EFSA Journal, 5*, 1–5.

European Medicines Agency. (2020). *HMA-EMA joint big data taskforce phase II report: 'Evolving data-driven regulation'* 1. www.ema.europa.eu

Guttinger, S. (2020). 'Frankenswine', genome editing, and the question of sustainable pig farming'. In *Interdisciplinarity, sustainability and expert disagreement*, online conference. https://interdisciplinarityandexpertdisagreement.wordpress.com/frankenswine-genome-editing-and-the-question-of-sustainable-pig-farming/

Guttinger, S. (2021). Process and practice: Understanding the nature of molecules. *HYLE: International Journal for Philosophy of Chemistry, 27*, 47–66.

Hansen, M., & Checker, M. (2016). *Hyde Park—A community's unrelenting pursuit for environmental justice.* www.hydeparkfilm.com

Leonelli, S. (2019). Philosophy of biology: The challenges of big data biology. *eLife, 8*, e47381.

Ornoy, A. (2009). Valproic acid in pregnancy: How much are we endangering the embryo and fetus? *Reproductive Toxicology, 28*(1), 1–10.

Perry, C. (2014). '*"Big data" Heralds a new kind of analyst*'. News and Events Blogpost. Harvard School of Engineering. https://seas.harvard.edu/news/2014/01/big-data-heralds-new-kind-ana lyst

Pharmafile. (2021). *How COVID-19 has changed pharmacovigilance*. Pharmafile.com. http://www. pharmafile.com/news/571507/how-covid-19-has-changed-pharmacovigilance. Published on 26 February 2021.

Rocca, E., & Andersen, F. (2017). How biological background assumptions influence scientific risk evaluation of stacked genetically modified plants: An analysis of research hypotheses and argumentations. *Life Sciences, Society and Policy, 13*, 11.

Rocca, E., & Anjum, R. L. (2019). Why causal evidencing of risk fails: An example from oil contamination. *Ethics, Policy & Environment, 22*, 197–213.

Rocca, E., & Grundmark, B. (2021). Monitoring the safety of medicines and vaccines in times of pandemic: Practical, conceptual, and ethical challenges in pharmacovigilance. *Argumenta, 781*, 127–146.

Rocca, E., Copeland, S., & Edwards, R. I. (2019). Pharmacovigilance as scientific discovery: An argument for trans-disciplinarity. *Drug Safety, 42*, 1115–1124.

The Lancet. (2004). The World Bank is finally embracing science. *The Lancet, 364*, 731–732.

Triviño Alonso, V., & Nuño de la Rosa, L. (2016). A causal dispositional account of fitness. *History and Philosophy of the Life Sciences, 38*, 1–18.

Triviño Alonso, V. (2020). Philosophical bias and adaptation to climate change: How fit is it, really? In *Interdisciplinarity, sustainability and expert disagreement*, online confer- ence. https://interdisciplinarityandexpertdisagreement.wordpress.com/philosophical-bias-and- adaptation-to-climate-change-how-fit-is-it-really/

Wickson, F., & Wynne, B. (2012). Ethics of science for policy in the environmental governance of biotechnology: MON810 maize in Europe. *Ethics, Policy & Environment, 15*, 321–340.

Concluding Remarks

Transforming Scientific Controversy into Constructive Dialogue

One of the greatest challenges today is how we can work together towards a sustainable future, across disciplines and perspectives. Training tomorrow's experts to identify, understand, and overcome some barriers for interdisciplinary collaboration and research is one strategy that we have proposed in this book. We hope to have shown that it's not sufficient that future professionals are well trained in their own areas of expertise, but that they also need to understand the perspectives of experts from other disciplines, and to recognise some implicit, foundational sources of interdisciplinary disagreement. This requires the joint effort of philosophers, scientists, and stakeholders of science.

Our aim has been to help prepare professionals to tackle problems together in inter- and trans-disciplinary teams, rather than separated by different disciplines and academic cultures. To think, evaluate, and act as a team requires that we understand different disciplinary perspectives and languages. While central scientific concepts are often used broadly and with multiple meanings, we have introduced a number of specific terms that are more transparent and name the specific philosophical bias. For instance, instead of talking generally about 'causality', one could say that a 'statistical difference-maker' has been identified, but no 'causal mechanism' to explain it. And rather than saying that an organism is 'complex', one can specify whether it is seen as 'composed' or 'emergent'. Although each of these concepts are themselves matters of philosophical debate, our idea here has been to introduce the curious reader to a different dimension of science, which can be studied more in-depth through the suggested further readings.

After reading this introduction to philosophy of science, we hope that you have acquired some tools for better understanding the non-empirical roots of scientific controversy and disagreement. We also hope to have explained why there can be divergent but equally rational evaluations of the same scientific evidence, and also

R. L. Anjum and E. Rocca, *Philosophy of Science*, Palgrave Philosophy Today, https://doi.org/10.1007/978-3-031-56049-1

why different disciplines might favour different methods and arguments. Finally, we hope you have learned how to recognise and explicate your own disciplinary and individual philosophical biases, and to communicate these in a transparent way.

If we have done our job, you should now be able to critically discuss the different possible interpretations of some basic concepts that are foundational for scientific knowledge, practice, and action: causality, probability, risk, inference, value, and complexity. You should also be able to relate different philosophical biases to norms and practices in your own discipline. Faced with a case of controversy, you can now use these tools to see if you can detect some non-empirical sources of disagreement, but this is not an easy task. The case analyses in Part III are meant to get you started, but also to illustrate how complex and multi-layered scientific controversies can be.

To conclude, we hope it won't go unnoticed that philosophy and critical reflection must play a central role in science moving forward into the post-normal era.

Bibliography

Andersen, F., Anjum, R. L., & Rocca, E. (2019). Philosophical bias is the one bias that science cannot avoid. *eLife, 8*, e44929.

Andersen, F., & Rocca, E. (2020). Underdetermination and evidence-based policy. *Studies in History and Philosophy of Science Part C: Studies in History and Philosophy of Biological and Biomedical Sciences, 84*, 101335.

Anderson, C. (2008). *The end of theory: The data deluge makes the scientific method obsolete.* WIRED. https://www.wired.com/2008/06/pb-theory/. Published on 23 June 2008.

Anderson, E. (2020). Feminist epistemology and philosophy of science. In E. N. Zalta (Ed.), *Stanford encyclopedia of philosophy* (Spring 2020 ed.). https://plato.stanford.edu/archives/spr2020/entries/feminism-epistemology

Anderson, G., Dupré, J., & Wakefield, J. G. (2019). Philosophy of biology: Drawing and the dynamic nature of living systems. *eLife, 8*, e46962.

Anjum, R. L., & Mumford, S. (2017). Emergence and demergence. In M. Paoletti & F. Orilia (Eds.), *Philosophical and scientific perspectives on downward causation* (pp. 92–109). Routledge.

Anjum, R. L., & Mumford, S. (2018). What probabilistic causation should be. In *Causation in science and the methods of scientific discovery* (pp. 165–173). Oxford University Press.

Anjum, R. L., & Rocca, E. (2019). From ideal to real risk: Philosophy of causation meets risk analysis. *Risk Analysis, 39*, 729–740.

Arendt, A. (1951). *The origins of totalitarianism.* Harcourt.

Arendt, A. (1967, February 25). Truth and politics. *The New Yorker.* https://www.newyorker.com/magazine/1967/02/25/truth-and-politics

Håkon Aspøy & Håkon Stokland (2022). Segmented forest realities: The ontological politics of biodiversity mapping. *Environmental Science & Policy, 137*, 120–127.

Aven, T. (2010). *Misconceptions of risk.* Wiley.

Bacon, F. (1620). *The new organon* (F. H. Anderson, Ed.). Bobbs-Merrill. A free version can be found on the Early Modern Texts webpage. https://www.earlymoderntexts.com/assets/pdfs/bacon1620.pdf

Beck, U. (1992). *Risk society: Towards a new modernity.* Sage.

Beebee, H. (2006). *Hume on causation.* Routledge.

Beebee, H., Hitchcock, C., & Menzies, P. (Eds.). (2009). *The Oxford handbook of causation.* Oxford University Press.

Bennett, C. M., Baird, A. A., Miller, M. B., & Wolford, G. L. (2010). Neural correlates of inter-species perspective taking in the post-mortem Atlantic salmon: An argument for proper multiple comparisons correction. *Journal of Serendipitous and Unexpected Results, 1*, 1–5.

Binimelis, R., & Wickson, F. (2019). The troubled relationship between GMOs and beekeeping: An exploration of socioeconomic impacts in Spain and Uruguay. *Agroecology and Sustainable Food Systems, 43*, 546–578.

Bjerkedal, T., Czeizel, A., Goujard, J., et al. (1982). Valproic acid and spina bifida. *The Lancet, 320*, 1096.

Bortolotti, L. (2008). *An introduction to the philosophy of science.* Polity Press.

Bouk, D. (2015). *How our days became numbered: Risk and the rise of the statistical individual.* University of Chicago Press.

Bourque, G., Burns, K. H., Gehring, M., et al. (2018). Ten things you should know about transposable elements. *Genome Biology, 19*, 199.

Bradford Hill, A. (1965). The environment and disease: Association or causation? *Proceeding of the Royal Society of Medicine, 58*, 295–300.

BREDL—Blue Ridge Environmental Defense League. (2014). *Waiting in a Cesspool.* https://bredl.org/resources/waiting-in-a-cesspool/

Bridgman, P. W. (1927). *The logic of modern physics.* Macmillan.

Browning, D., & Myers, W. T. (Eds.). (1998). *Philosophers of process.* Fordham.

Buehler, J. (2021, September 1). The complex truth about "junk DNA". *Quanta Magazine.* https://www.quantamagazine.org/the-complex-truth-about-junk-dna-20210901/#

Buolamwini, J., & Gebru, T. (2018). Gender shades: Intersectional accuracy disparities in commercial gender classification. *Proceedings of Machine Learning Research, 81*, 1–15.

Cambridge Elements in Philosophy of Science. Cambridge University Press. https://www.cambridge.org/core/publications/elements/philosophy-of-science

Cartwright, N. (1983). *How the laws of physics lie.* Clarendon Press.

Cartwright, N. (1999). *The dappled world: A study of the boundaries of science.* Cambridge University Press.

Cartwright, N. (2007). *Hunting causes and using them: Approaches in philosophy and economics.* Cambridge University Press.

Cartwright, N. (2012). Will this policy work for you? Predicting effectiveness better: How philosophy helps. *Philosophy of Science, 79*, 973–989.

Cartwright, N., Psillos, S., & Chang, H. (2003). Theories of scientific method. In M. J. Nye (Ed.), *The Cambridge history of science* (vol. 5, pp. 21–35). *Modern physical and mathematical sciences.* Cambridge University Press.

Catalogue of Bias (D. Nunan & C. Heneghan, Eds.). Centre for Evidence-Based Medicine. University of Oxford. https://catalogofbias.org/

Chalmers, A. F. (2013). *What is this thing called science?* Hackett Publishing.

Checker, M. (2007). "But i know it's true": Environmental risk assessment, justice, and anthropology. *Human Organization, 66*, 112–124.

Collins, J., Hall, N., & Paul, L. A. (Eds.). (2004). *Causation and counterfactuals.* MIT Press.

Copeland, S. (2017). On serendipity in science: Discovery at the intersection of chance and wisdom. *Synthese, 196*, 2385–2406.

Crasnow, S., & Intemann, K. (Eds.). (2021). *The Routledge handbook of feminist philosophy of science.* Routledge.

Crenshaw, K. (1989). Demarginalizing the intersection of race and sex: A black feminist critique of antidiscrimination doctrine, feminist theory and antiracist politics. *University of Chicago Legal Forum, 140*, 139–167.

Criado Perez, C. (2019). *Invisible women: Data bias in a world designed for men.* Abrams Press.

Criado Perez, C. (Podcast). *Visible women.* Tortoise Media. https://www.tortoisemedia.com/listen/visible-women/

Curd, M., & Psillos, S. (Eds.). (2014). *The Routledge companion to philosophy of science.* Routledge.

D'Ignazio, C., & Klein, L. F. (2020). *Data feminism.* The MIT Press.

Dally, A. (1998). Thalidomide: Was the tragedy preventable? *The Lancet, 351*, 1197–1199.

Dawkins, R. (1976). *The selfish gene.* Oxford University Press.

Dellsén, F., & Baghramian, M. (Eds.). (2021, November). Special Issue on disagreement in science. *Synthese, 198*(25), 6011–6021.

Dennett, D. (2013). *Intuition pumps and other tools for thinking*. W. W. Norton.

DiLiberti, J. H., Farndon, P. A., Dennis, N. R., & Curry, C. J. (1984). The fetal valproate syndrome. *American Journal of Medical Genetics Part A, 19*, 473–481.

Directorate General for Health and Consumers Evaluation of the EU. (2010). *Evaluation of the EU Legislative Framework in the Field of Medicated Feed, Final Report*. https://food.ec.europa.eu/system/files/2016-10/gmo_rep-stud_2010_report_eval-gm.pdf

Douglas, H. (2000). Inductive risk and values in science. *Philosophy of Science, 67*, 559–579.

Douglas, H. (2012). Weighing complex evidence in a democratic society. *Kennedy Institute of Ethics Journal, 22*, 139–162.

Dowe, P. (2000). *Physical causation*. Cambridge University Press.

Dupré, J., & Guttinger, S. (2016). Viruses as living processes. *Studies in History and Philosophy of Science Part C: Studies in History and Philosophy of Biological and Biomedical Sciences, 59*, 109–116.

Epilim Summary of Product Characteristics. https://www.medicines.org.uk/emc/product/519/smpc#gref

European Food and Safety Agency (EFSA). (2007). Guidance document for the risk assessment of genetically modified plants containing stacked transformation events by the Scientific Panel on Genetically Modified Organisms (GMO). *EFSA Journal, 5*, 1–5.

European Medicines Agency. (2020). *HMA-EMA joint big data taskforce phase II report: 'Evolving data-driven regulation'* 1. www.ema.europa.eu

Ewen, S. W., & Pusztai, A. (1999). Effect of diets containing genetically modified potatoes expressing galanthus nivalis lectin on rat small intestine. *The Lancet, 354*, 1353–1354.

Feyerabend, P. (1975a). How to defend society against science. *Radical philosophy* (pp. 261–271). Stegosaurus Press.

Feyerabend, P. (1975b). *Against method*. New Left Books.

Funtowicz, S., & Ravetz, J. (1993). Science for the post-normal age. *Futures, 25*, 739–755.

Giddens, A. (1999). Risk and responsibility. *The Modern Law Review, 62*, 1–10.

Gillies, D. (2000). Varieties of propensities. *The British Journal for Philosophy of Science, 51*, 807–835.

Gillies, D. (2018). *Causality, probability, and medicine*. Routledge.

Gower, B. (1997). *Scientific method: An historical and philosophical introduction*. Routledge.

Guttinger, S. (2020). 'Frankenswine', genome editing, and the question of sustainable pig farming'. In *Interdisciplinarity, sustainability and expert disagreement*, online conference. https://interdisciplinarityandexpertdisagreement.wordpress.com/frankenswine-genome-editing-and-the-question-of-sustainable-pig-farming/

Guttinger, S. (2021). Process and practice: Understanding the nature of molecules. *HYLE: International Journal for Philosophy of Chemistry, 27*, 47–66.

Haack, S. (1996). Science as social? Yes and no. In L. H. Nelson & J. Nelson (Eds.), *Feminism, science, and the philosophy of science*, Synthese Library (Studies in epistemology, logic, methodology, and philosophy of science, vol. 256). Springer.

Haack, S. (2001). Clues to the puzzle of scientific evidence. *Principia: An International Journal of Epistemology, 5*, 253–281.

Haack, S. (2002). The same, only different. *Journal of Aesthetic Education, 36*, 34–39.

Han, H., & Jain, A. K. (2014). Age, gender and race estimation from unconstrained face images. *Michigan State University Technical Report, 14–5*, 1–9.

Hansen, M., & Checker, M. (2016). *Hyde Park—A community's unrelenting pursuit for environmental justice*. www.hydeparkfilm.com

Hanson, N. R. (1958). *Patterns of discovery an inquiry into the conceptual foundations of science*. Cambridge University Press.

Haraway, D. (1988). Situated knowledges: The science question in feminism and the privilege of partial perspective. *Feminist Studies, 14*, 575–599.

Harding, S. (1991). *Whose science? Whose knowledge? Thinking from women's lives.* Cornell University Press.

Harding, S. (Podcast interview 2016, September 7). *Interviewed by Emily Crandall for New Books Network podcast.* https://newbooksnetwork.com/sandra-harding-objectivity-and-diversity-a-new-logic-of-scientific-inquiry-u-of-chicago-press-2015

Hawking, S., & Mlodinow, L. (2010). *The grand design.* Penguin Random House.

Henrich, J., Heine, S. J., & Norenzayan, A. (2010). The weirdest people in the world? *Behavioral and Brain Sciences, 33,* 61–83.

Hepburn, B., & Andersen, H. (2021). Scientific method. *Stanford encyclopedia of philosophy* (Summer 2021 ed.). E. N. Zalta (Ed.). https://plato.stanford.edu/archives/sum2021/entries/scientific-method

Hesse, M. (1974). *The structure of scientific inference.* Macmillan.

Howick, J. (2011). *The philosophy of evidence-based medicine.* Wiley-Blackwell.

Hume, D. (1739). *A treatise of human nature* (L. A. Selby-Bigge, Ed., 1888th ed.). Clarendon Press.

Hume, D. (1748). *An enquiry concerning human understanding* (L. Selby-Bigge, Ed.). Clarendon Press.

Hurtig, A.-K., & San Sebastián, M. (2002). Geographical differences in cancer incidence in the Amazon basin of Ecuador in relation to residence near oil fields. *International Journal of Epidemiology, 31,* 1021–1027.

Hustwit, J. R. (Online entry). *Process philosophy: Internet encyclopedia of philosophy.* https://iep.utm.edu/processp/

ICRP. (1975). *Report of the task group on reference man.* ICRP Publication 23. Pergamon Press.

Illari, P., & Russo, F. (2014). *Causality: Philosophical theory meets scientific practice.* Oxford University Press.

Illari, P., & Russo, F. (Eds.). (2024). *Routledge handbook of causality and causal methods.* Routledge.

Illary, P., Russo, F., & Williamson, J. (Eds.). (2011). *Causality in the sciences.* Oxford University Press.

Ingthorsson, R. D. (2013). The natural vs. human sciences: Myth, methodology, and ontology. *Discusiones Filosóficas, 14,* 25–41.

Ingthorsson, R. D. (2021). *A powerful particulars view of causation.* Routledge.

Interdisciplinarity, Sustainability and Expert Disagreement. (2020). Online conference. https://interdisciplinarityandexpertdisagreement.wordpress.com

Ioannidis, J. P. A. (2005). Why most published research findings are false. *PLoS Medicine, 2,* e124. https://doi.org/10.1371/journal.pmed.0020124

Ivanova, M. (2021). *Duhem and holism.* Cambridge University Press.

Kao, J., Brown, N. A., Schmid, B., et al. (1981). Teratogenicity of valproic acid: In vivo and in vitro investigations. *Teratogenesis, Carcinogenesis and Mutagenesis, 1,* 367–382.

Kuhn, T. (1962). *The structure of scientific revolutions.* University of Chicago Press.

Kuhn, T. (1977). *Objectivity, value judgment and theory choice.* University of Chicago Press.

Kusch, M. (2017). *'Relativism', 5 short online lectures.* University of Edinburgh. https://youtube.com/playlist?list=PLKuMaHOvHA4o6JF3O-8mBeUjsT0Hv1DIY

Kusch, M. (2020). *Relativism in the philosophy of science.* Cambridge Elements Philosophy of Science Series.

Lakatos, I. (1973). *Science and pseudoscience.* LSE podcast and transcript. https://www.lse.ac.uk/philosophy/science-and-pseudoscience-overview-and-transcript/

Lakatos, I. (1978). *The methodology of scientific research programmes* (Philosophical papers: Volume 1, J. Worrall & G. Currie, Eds.). Cambridge University Press.

Leonelli, S. (2016). *Data-centric biology: A philosophical study.* University of Chicago Press.

Leonelli, S. (2019). Philosophy of biology: The challenges of big data biology. *eLife, 8,* e47381.

Leonelli, S. (2023). *Philosophy of open science.* Elements in the Philosophy of Science. Cambridge University Press.

Lie, S. A. N. (2017). *Philosophy of nature: Rethinking naturalness.* Routledge.

Lindquist, M. (2008). VigiBase, the WHO global ICSR database system: Basic facts. *Drug Information Journal, 42*, 409–419.

Longino, H. (1990). *Science as social knowledge: Values and objectivity in scientific inquiry.* Princeton University Press.

Longino, H. (Online Lecture, 2021). *Critical contextual empiricism, diversity and inclusiveness.* Weizsäcker-Zentrum Universität Tübingen. YouTube. https://www.youtube.com/watch?v=Xys Yymrh7IE

Machamer, P., Darden, L., & Craver, C. F. (2000). Thinking about mechanisms. *Philosophy of Science, 67*, 1–25.

Massimi, M. (Ed.). (2015). *Philosophy and the sciences for everyone.* Routledge.

Massimi, M. (2022). *Perspectival realism.* Oxford University Press.

Massimi, M. et al. (Free online course). *Philosophy and the sciences: Examining philosophy's relationship with the physical and cognitive sciences.* https://www.ed.ac.uk/ppls/philosophy/res earch/impact/free-online-courses/philosophy-and-the-sciences

McNeish, J., Borchgrevink, A., & Logan, O. (Eds.). (2015). *Contested powers: The politics of energy and development in Latin America.* Zed Books.

Mellor, D. H. (1971). *The matter of chance.* Cambridge University Press.

Mellor, D. H. (2002). *The facts of causation.* Routledge.

Muggli, M. E., Forster, J. L., Hurt, R. D., & Repace, J. L. (2001). The smoke you don't see: Uncovering tobacco industry scientific strategies aimed against environmental tobacco smoke policies. *American Journal of Public Health, 91*, 1419–1423.

Mumford, S., & Anjum, R. L. (2011). *Getting causes from powers.* Oxford University Press.

Mumford, S., & Anjum, R. L. (2013). *Causation: A very short introduction.* Oxford University Press.

Nicholson, D. J., & Dupré, J. (2018). *Everything flows: Towards a processual philosophy of biology.* Oxford University Press.

No Patents of Seeds. https://www.no-patents-on-seeds.org/

Okasha, S. (2002). *Philosophy of science: A very short introduction.* Oxford University Press.

Ornoy, A. (2009). Valproic acid in pregnancy: How much are we endangering the embryo and fetus? *Reproductive Toxicology, 28*(1), 1–10.

Oughton, D. (2003). Protection of the environment from ionising radiation: Ethical issues. *Journal of Environmental Radioactivity, 66*, 3–18.

Paul, L. A., & Hall, N. (2013). *Causation: A user's guide.* Oxford University Press.

Perry, C. (2014). '*"Big data" Heralds a new kind of analyst'.* News and Events Blogpost. Harvard School of Engineering. https://seas.harvard.edu/news/2014/01/big-data-heralds-new-kind-ana lyst

Peterson, C. H., Rice, S. D., Short, J. W., et al. (2003). Long-term ecosystem response to the Exxon Valdez oil spill. *Science, 302*, 2082–2086.

Pharmafile. (2021). *How COVID-19 has changed pharmacovigilance.* Pharmafile.com. http://www. pharmafile.com/news/571507/how-covid-19-has-changed-pharmacovigilance. Published on 26 February 2021.

Pietryska, J. (2017). *Looking forward: Prediction and uncertainty in modern America.* University of Chicago Press.

Popper, K. (1959). The propensity interpretation of probability. *British Journal of Philosophy of Science, 10*, 25–42.

Popper, K. (1990). *A world of propensities.* Thoemmes.

Portier, C. J., Armstrong, B. K., Baguley, B. C., et al. (2016). Differences in the carcinogenic evaluation of glyphosate between the International Agency for Research on Cancer (IARC) and the European Food Safety Authority (EFSA). *The Journal of Epidemiology & Community Health, 70*, 741–745.

Pray, L., & Zhaurova, K. (2008). Barbara McClintock and the discovery of jumping genes (Transposons). *Nature Education, 1*, 169. https://www.nature.com/scitable/topicpage/barbara-mcclin tock-and-the-discovery-of-jumping-34083/

Pray, L. A. (2008). Discovery of DNA structure and function: Watson and Crick. *Nature Education, 1*, 100.

Ramsey, F. P. (1926). Truth and probability. In F. P. Ramsey (Ed.), *The foundations of mathematics and other logic essays* (pp. 156–198). Routledge, 1931.

Ravindran, S. (2012). Barbara McClintock and the discovery of jumping genes. *Proceedings of the National Academy of Sciences, 109*, 20198–20199.

Reiss, J. (2015). *Causation, evidence, and inference.* Routledge.

Rescher, N. (1996). *Process metaphysics: An introduction to process philosophy.* SUNY Press.

Riedel, S. (2005). Edward Jenner and the history of smallpox and vaccination. *Proceedings* (Baylor University, Medical Center*), 18*, 21–25.

Rocca, E. (2020a). Philosophy of science meets patient safety. *Uppsala Reports, 82*, 16–19.

Rocca, E. (Podcast 2020b). Scientific disagreement and philosophy. *PedPod—NMBU Pedagogy Podcast*, Episode 4. https://www.nmbu.no/ansatt/laringssenteret/kurs-og-kompetanse/pedpod

Rocca, E., & Andersen, F. (2017). How biological background assumptions influence scientific risk evaluation of stacked genetically modified plants: An analysis of research hypotheses and argumentations. *Life Sciences, Society and Policy, 13*, 11.

Rocca, E., & Anjum, R. L. (2019). Why causal evidencing of risk fails: An example from oil contamination. *Ethics, Policy & Environment, 22*, 197–213.

Rocca, E., Copeland, S., & Edwards, R. I. (2019). Pharmacovigilance as scientific discovery: An argument for trans-disciplinarity. *Drug Safety, 42*, 1115–1124.

Rocca, E., & Grundmark, B. (2021). Monitoring the safety of medicines and vaccines in times of pandemic: Practical, conceptual, and ethical challenges in pharmacovigilance. *Argumenta, 781*, 127–146.

Rudner, R. (1953). The scientist qua scientist makes value judgments. *Philosophy of Science, 20*, 1–6.

Sackett, D. (1979). Bias in analytic research. *Journal of Chronic Diseases, 32*, 51–63.

Sanches de Oliveira, G., & Baggs, E. (2023). *Psychology's WEIRD problems* (Elements in psychology and culture). Cambridge University Press.

Schardein, J. (1985). *Chemically induced birth defects.* Marcel Dekker.

Schick, S. F., & Glantz, S. A. (2007). Old ways, new means: Tobacco industry funding of academic and private sector scientists since the master settlement agreement. *Tobacco Control, 16*, 157–164.

SDHS Research Report. (2018). *Status of patenting plants in the global south.* https://sdhsprogram.org/document/statusofpatentingplantsintheglobalsouth/

Seed: The Untold Story. (2016). https://www.seedthemovie.com/

Seibt, J. (2012/2017). Process philosophy. *Stanford encyclopedia of philosophy* (Summer 2023 ed.). E. N. Zalta & U. Nodelman (Eds.). https://plato.stanford.edu/archives/sum2023/entries/process-philosophy/

Siemiatycki, J. (2002). Commentary: Epidemiology on the side of the angels. *International Journal of Epidemiology, 31*, 1027–1029.

Sonnenschein. C., & Soto, A. M. (1999). *The society of cells: Cancer and control of cell proliferation.* Springer Verlag.

Staley, K. (2014). *An introduction to the philosophy of science.* Cambridge University Press.

Suárez, M. (2021). *Philosophy of probability and statistical modelling.* Elements in the philosophy of science. Cambridge University Press.

Sulston, J., & Ferry, G. (2002). *The common thread: A story of science, politics, ethics and the human genome.* Joseph Henry Press.

Tbakhi, A., & Amr, S. S. (2007). Ibn Al-Haytham: Father of modern optics. *Annals of Saudi Medicine, 27*, 464–467. https://www.ncbi.nlm.nih.gov/pmc/articles/PMC6074172/

The Human Genome Project. National Human Genome Institute. https://www.genome.gov/human-genome-project

The Lancet. (2004). The World Bank is finally embracing science. *The Lancet, 364*, 731–732.

Triviño Alonso, V. (2020). Philosophical bias and adaptation to climate change: How fit is it, really? In *Interdisciplinarity, sustainability and expert disagreement*, online conference. https://interdisciplinarityandexpertdisagreement.wordpress.com/philosophical-bias-and-adaptation-to-climate-change-how-fit-is-it-really/

Triviño Alonso, V., & Nuño de la Rosa, L. (2016). A causal dispositional account of fitness. *History and Philosophy of the Life Sciences, 38*, 1–18.

Tuntreet. (2020). Open letter: A call for promoting critical thinking for interdisciplinarity in NMBU. *Tuntreet* (NMBU student newspaper), *6*(75). https://issuu.com/tuntreet. Published on 10 September 2020.

Ward, L. (2022). *The empress and the English doctor how Catherine the great defied a deadly virus.* Blackwell.

Weinberger, N., & Bradley, S. (2020). Making sense of non-factual disagreement in science. *Studies in History and Philosophy of Science Part A, 83*, 36–43.

Whitehead, A. F. (1929). *Process and reality.* The Free Press.

Wickson, F., & Wynne, B. (2012a). The anglerfish deception: The light of proposed reform in the regulation of GM crops hides underlying problems in EU science and governance. *EMBO Reports, 13*, 100–105.

Wickson, F., & Wynne, B. (2012b). Ethics of science for policy in the environmental governance of biotechnology: MON810 maize in Europe. *Ethics, Policy & Environment, 15*, 321–340.

Wilhelm, J. P. (2020). *Ethiopian teff: The fight against biopiracy.* DW.com. https://www.dw.com/en/ethiopian-teff-the-fight-against-biopiracy/a-52085081. Published on 21 January 2020.

Woodward, J. (2003). *Making things happen: A theory of causal explanation.* Oxford University Press.

World Bank. (2005). *Project performance assessment report, Bangladesh nutrition project* (Report No. 32563). https://ieg.worldbankgroup.org/sites/default/files/Data/reports/ppar_3 2563.pdf. Published on 13 June 2005.

Xiao, X., et al. (2020). A genetically defined compartmentalized striatal direct pathway for negative reinforcement. *Cell, 183*, 211–227.

Vogel, F. (1995). Widukind Lenz. *European Journal of Human Genetics, 3*, 384–387. https://doi.org/10.1159/000472329

Von Mises, R. (1928). *Probability, statistics and truth* (2nd revised English ed.). Allen & Unwin. 1961.